FATNESS AND THE MATERNAL BODY

Fertility, Reproduction and Sexuality

GENERAL EDITORS:
David Parkin, Fellow of All Souls College, University of Oxford.
Soraya Tremayne, Founding Director, Fertility and Reproduction Studies Group and Research Associate, Institute of Social and Cultural Anthropology, University of Oxford.
Marcia C. Inhorn, William K. Lanman, Jr. Professor of Anthropology and International Affairs, Yale University.
Philip Kreager, Director, Fertility and Reproduction Studies Group, and Research Associate, Institute of Social and Cultural Anthropology and Institute of Human Sciences, University of Oxford.

FATNESS AND
THE MATERNAL BODY

Women's Experiences of Corporeality
and the Shaping of Social Policy

Edited by
Maya Unnithan-Kumar
and Soraya Tremayne

Berghahn Books
New York • Oxford

First published in 2011 by

Berghahn Books

www.BerghahnBooks.com

© 2011 Maya Unnithan-Kumar and Soraya Tremayne

Library of Congress Cataloging-in-Publication Data

Fatness and the maternal body : women's experiences of corporeality
and the shaping of social policy / edited by Maya Unnithan-Kumar
and Soraya Tremayne.
    p. cm. -- (Fertility, reproduction and sexuality)
  Includes bibliographical references and index.
  ISBN 978-0-85745-122-4 (hardback : alk. paper) —
  ISBN 978-0-85745-123-1 (ebook)
  1. Obesity in women. 2. Women—Physiology. 3. Body image in
women. 4. Human body—Social aspects. 5. Human body—Symbolic
aspects. I. Unnithan-Kumar, Maya, 1961– II. Tremayne, Soraya.
  RA625.O23F37 2011
  362.196'3980082--dc23

                                                      2011017212

British Library Cataloguing in Publication Data

A catalogue record for this book is available from the British Library.

Printed in the United States on acid-free paper

ISBN 978-0-85745-122-4 (hardback)
E-ISBN 978-0-85745-123-1

CONTENTS

LIST OF FIGURES

LIST OF TABLES

PREFACE

This volume is the result of a workshop and a seminar series
organised by the Fertility and Reproduction Studies Group
(FRSG), at the Institute of Social and Cultural Anthropology,
University of Oxford during the autumn term 2006. The focus of
both the workshop and seminars was on body weight, predominantly
'fatness', and its effects on reproduction. The contributors came
from disciplines as varied as anthropology, sociology, public health,
demography, and psychology.

The book as a whole explores the meaning of being overweight
and its links with reproduction from an interdisciplinary and cross-
cultural perspective. It also examines the interaction between state
policies to avert the health risks involved in being obese and the
individual's responses to these. Statistics demonstrate an alarming
increase in obesity, which is now recognised as a major health risk
at a global scale. A considerable amount of research continues to be
carried out to establish the causes and consequences of being
overweight. The effects of fatness on women's reproductive health,
especially during the childbearing years, feature prominently across
these studies.

It was with this in mind that the FRSG thought it timely to
organise a workshop to explore the facts and assumptions related to
obesity and reproduction. In the volume, the term 'obesity' is used
when referring to the biomedical sense of being overweight while
'fatness' is used when conveying the more social and cultural
understanding of the condition. This is not to deny the fact that
obesity itself is culturally constructed.

Discussions of how and why being overweight is considered
unhealthy, unsightly, costly for the state, a sign of poverty and
uneducated, have all arisen in current debates on the subject,
especially in the well-resourced health care settings of post-industrial

countries. Specialists from various disciplines continue to inform the public on the risks involved in being 'fat'. The media provides a broad coverage on the harmful aspects of being overweight, which includes health, fashion, leisure and life-style. Health planners and medical advisers organise special clinics and advice on life-styles, and demand for their services is on the increase. Food manufacturing industries have made major alterations to their products, and are required to provide information on ingredients for the consumers. Furthermore, biomedical research has shown the risks involved to the children born of fat women, including a predisposition to becoming fat themselves, and in turn passing it on to their children. In short, the picture looks alarming all round and there seems to be a general air of panic about growing body weight. Yet, despite health warnings and a heightened public awareness, the number of obese people remains steadily on the increase. This has spurred on other kinds of questions as to what is fatness, how people perceive their bodies, what are their own interpretations of their body weight, whether the involvement of the state to devise policies to counter the negative effects of fatness on individuals and society has other unintended consequences, and whether nutritional intervention is necessarily effective?

One of the general assumptions has been that the modern lifestyle of post-industrial countries is responsible for an increase in fatness, whose example has been followed by other countries. However, the chapters in this volume throw a different light on the question of body weight and challenge the rather simplistic suggestion that fatness stems from biomedical factors alone. The question of the ideal size body is a complex and multi-faceted one which is variously defined by the dynamics of the relationship between the individual, the community and the environment. The organisers of the workshop aim at presenting some of the less well-understood and deeply rooted motivations which determine the control of the body weight in different social, economic and cultural contexts.

The complexities involved point to the differences within the same culture, to class, gender, generation, and to the responses of the individual vis-à-vis the actions of state policies. This volume therefore reflects the complexities involved in tackling, perceiving and defining what a fat body is in its relation to reproduction.

The Editors are aware that the focus of this volume is on women and their reproductive health in relation to their body weight, and that there is no discussion of male fatness and its consequences on their reproductive life. Links between male fatness and reproduction are also significant, and this remains a seriously under-researched area. But there is interesting work emerging in this area which is not confined to the countries of the North.

We thank all the contributors to the volume for their enthusiasm, hard work and patience. The Editors are particularly grateful to Renate Barber, who contributed an overview paper which opened the discussion at the workshop. Renate also summarised the papers for a concluding discussion. Her observations have been helpful to all the participants and are embedded throughout the text. Finally, we owe special thanks to Catriona Shepard for her valuable editorial work on the volume, and to Ammara Maqsood.

Soraya Tremayne
Oxford, April 2011

Chapter 1

INTRODUCTION

CORPOREALITY AND REPRODUCTION:

UNDERSTANDING FATNESS THROUGH THE DIVERSE EXPERIENCES OF MOTHERHOOD, CONSUMPTION AND SOCIAL REGULATION

Maya Unnithan-Kumar

S tarting from the premise that the body is socially, culturally and historically constituted, this volume examines how maternal bodies are made meaningful through the discourse of gender, size and reproduction. In particular, corporeality and what it means to be fat or thin is explored in relation to women's physical and subjective experiences of fertility, pregnancy, childbirth and lactation. The body has increasingly become a central means in the social sciences of understanding how power is exercised, social identity and inequality experienced.[1] The fat body itself has become the focus of diverse and contested debates within the social, medical and health sciences (for a comprehensive view see Gremillion 2005; Gard and Wright 2005; Jutel 2006; Throsby 2007; Monagahan 2008; and Gilman 2008, among others). Drawing on some of this literature, the main focus of the current volume is on fatness, motherhood and maternal obesity.[2] How does the shape and size of the maternal body affect women's conceptions of self and their engagement with others? In what ways is the maternal physical body lived and experienced?

Experiences of maternity are being shaped by the language and ideas of biomedicine and public health globally. Especially within better-resourced healthcare settings there is increasing attention being paid to the risks related to maternal obesity (e.g., the U.K. National Audit Office 2001). Large women, according to these health policy guidelines, need medical and health scrutiny because of the maternal and childbearing risks that being overweight pose for them. In biomedical perspectives the increasing prevalence of the 'overweight' body is an emerging pathological human condition connected to modern lifestyles where 'obesogenic environments' provide an easy access to high density foods at the same time as the opportunities and/or inclination to exercise are reduced (NAO 2001). As Jutel suggests, from being a sign or a symptom, being overweight is becoming a disease entity evident in the frequency and way in which the term is being used in the media, the medical establishment and by the public (2006). It is not only the power to characterise a condition as a disease but equally the authority that such labelling exerts on lay perceptions, making it seem 'natural' (Jordan, 1997) and in permeating everyday ways of perceiving the body and self (Lock 2001). One of the aims of the present volume is to investigate the extent to which medical and health perspectives on maternal fatness carry authority among women in different social contexts. What does it mean to think about the body shape of women as mothers in terms of 'risk' and of 'fitness' as a category in opposition to 'fatness'? How might public health initiatives address an audience that do not identify with the 'problem'? A focus on fatness in the context of human reproduction and motherhood offers further instructive insights into the global circulation and authority of biomedical facts on fatness (as obesity, as risk, as anti-fitness, for example).

Like other social and cultural studies critical of the clinical and health policy discourse on obesity (for example, Martin 1987; Sontag 1989; Bordo 2003; Kulick and Meneley 2005; Monaghan 2008; Gard and Wright 2005) this volume challenges the spontaneous connection being made in scientific and popular discourse between fatness and ill health. Through a focus on cross-cultural studies on childbearing as well as bio-medically orientated critiques of maternal health, the volume aims to further investigate the social and cultural aspects of 'fat phobia'. The use of cross-regional, comparative accounts of fatness highlights that it is not easy to distinguish between contexts where fatness is celebrated and those where it is feared, as popular and somewhat romanticised understandings have us believe. Ideas of fatness exist in resource-poor countries where fatness may be both an appreciated or stigmatised condition associated with reproduction (as accounts from central Tanzania and from the Sahel in this volume demonstrate). We do not use cross-cultural comparisons as a means of reinforcing the boundaries between the 'West' and 'the rest' fully aware that these kind of juxtapositions are themselves a product of Eurocentric thinking on the body (Gremillion 2005). Contributions in this volume show instead the complexity that surrounds understandings of fatness and how they mainly stem from the diverse ways in which maternal corporeality is socially embedded across the divisions of culture, class, ethnicity, religion, gender and sexuality. The analytic use of such a comparative approach is that it enables one to 'see' (know) why what may be taken for granted in one context becomes the focus of attention in another. The fact that notions of fatness as connected to human reproduction are so dissimilar across time and space reinforces the significance of perceptions, experiences and practices, in short an embodied perspective, in understanding fatness. As a number of developing countries modernise and go through what public health scientists are calling 'the nutritional transition' such a grounded perspective also tells us how values and ideas about fatness and maternal identity are themselves transformed in the process of their circulation.

While the subjective accounts of fatness in the volume act as a counter to the predominantly objective ways in which the issue of body size has been treated in health policy literature, other chapters engage much more closely with objective understandings of the body. These contributions which are less subjectively based provide a more familiar (i.e., scientific) account of fatness in human reproduction at the same time as they provide a unique perspective on maternal fatness. As I discuss in the section below on Food, Fatness and Fertility, for nutritional anthropologists and reproductive biologists in particular, fat has a specific relation to female fertility and the onset of menarche. Rather than excluding biological perspectives, we find that these accounts provide a good standpoint to reflect on the strengths and limitations of the different disciplinary approaches to studying maternal body size as well as to the ways in which the 'work' of culture is construed: as supportive of, or detrimental to, maternal and child health. Subjectively inclined contributions, on the other hand, examine how the discourse around fatness and thinness may be experienced at the individual level, mediated through personal circumstance and resisted, accepted or ignored in daily life. The extent to which people choose between, or integrate, competing forms of knowledge (experiential versus scientific) in their quest to be healthy and accepted members of the society they live in, is an issue addressed by these contributions in the volume.

An important subtheme to the volume is a consideration of motherhood as a physical state and how this is connected with the moral valuations of women as social persons. One of the inevitable challenges associated with a focus on motherhood is how to avoid the pitfall of constructing women's identity as based primarily on their reproductive roles. Feminist scholars tend to regard the emphasis on women's reproductive roles as reinforcing an essentialised view of women as primarily determined by their childbearing. This discomfort with an emphasis on the experiential significance of pregnancy and birth, as Bordo points out, stems from an attendant 'fear of the conceptual proximity of such notions to constructions of mothering as the true destiny for women' (2003: 95). It is therefore important at the outset of the present volume to point out that it is precisely through a close and comparative reading of women's bodies as experienced through their size that we are able to examine the relationship between the sexual and maternal aspects of a woman's identity.

There are four main themes which arise from the material presented in the chapters and summarised below. The themes situate understandings of maternal fatness within: (1) culturally framed notions of the biological (Nature); (2) gender and sexual inequalities; (3) food and nutritional perspectives; and (4) global notions of risk and modernity.

Rethinking nature through fatness

One of the key theoretical concerns in the volume is the 'traffic' in ideas about nature and culture and the different ways in which these ideas are connected through concepts and practices to do with the body. This is far from a new theoretical concern (e.g., Haraway 1993; Bordo 1993; Martin 1987, 1994; Franklin 1995; Strathern 1972). But what is perhaps distinctive to the volume is an examination of the ways in which biology works within and through cultural understandings of the reproductive body to provide new analytic insight into fatness. Human birthing and parenting are particularly pertinent fields to think about the relationship between the biological and the social (Ginsburg and Rapp 1995) and what is considered 'natural' in human society, including the unequal relation between men and women as based on their biological differences, as feminist scholarship has shown (e.g., Bordo 2003). Feminist work has long been important for situating an understanding of gender inequality within cultural ideas about nature. Cross-cultural work in this stream has challenged the dichotomous relationship between the concepts of nature and culture as a way of thinking about gender inequality as also the assumption that the body is viewed through a universal biological lens (MacCormack and Strathern 1980; Strathern 1972). Women's visibly closer physical association with childbearing, popularly used to legitimise their gender difference and naturalise their inequality (Ortner 1974) has similarly been challenged. The idea that nature is itself a cultural category whose meanings vary was highlighted in these early works. Technological and genetic interventions in assisting reproduction have further reinforced the idea that biology itself is contingent rather than universal and recognised in ways that make it more social rather than 'natural' (Franklin 2001; Rose 2007).

Strathern and Franklin's critique of nature can also be used to destabilise universal biological understandings of fatness, connecting it instead to social and cultural interpretation, processes and valuations. Fat mothers' bodies which are doubly grounded in biology, i.e. as reproductive and as fat, then have to be reconstituted through cultural and gender frameworks. In contexts where biology is itself a social fact, we need to pay attention to the ways in which fat pregnant women's double 'grounding' affects their identity and perceptions of the self. In especially fat-phobic societies, what is the cultural work required by large-bodied pregnant women to combat the stigma attached to their condition? Throsby's work gives some idea of the kinds of responses we might expect to find. Her U.K.-based study on weight loss surgery patients documents, for example, three kinds of discourse operative, which challenge fatness as individual moral failure: the notion of fat-prone bodies (predisposed through genes or metabolism), childhood weight gain and life events disrupting the individuals' capabilities for weight management (2007).

The everyday context in which fat maternal bodies are made meaningful, and connected to local understandings of nature forms the core of the chapter by Warin, Moore and Davies in this volume. Drawing on Franklin and Haraways's metaphor of 'trafficking nature', Warin et al. analyse how a set of women identified as clinically obese in Adelaide draw upon a selective understanding of biology to positively evaluate their bodies, thus distancing themselves from the clinical classification of their condition. It is through a description of themselves as 'large boned' rather than fat that the women are able to establish a specific kinship (where large bones signify a shared family condition) at the same time as they deflect the stigma associated with being fat. The women in Warin et al.'s study also draw on their positive reproductive and nurturing attributes to explain their large bodies where superior quality is associated with such physical conditions as 'wide childbearing hips' and 'big shoulders'.

The analysis of fatness and thinness in illuminating the interconnections between the biological and the social is a theme further elaborated upon in Mabilia's chapter on childbearing and breastfeeding among the Gogo in Tanzania. The birth of 'good fat' babies is the result of the appropriate sexual and social conduct of the parents enabling the proper flow of bodily fluids between mother, father and child. The appropriate social and sexual behaviour of the woman during pregnancy is confirmed through the quality of her breast milk. Breast milk can be fat producing (good milk) as well as fat destroying (spoilt milk). Here the biological outcomes are contingent upon and constituted through social relations. It becomes important then to examine the social, relational context in which reproduction is itself located.

Reproducing the fat body through social relations: gender, sex, kinship, class

In her early work on the body as a metaphor or symbol of society, Douglas (1979) gives us an understanding of how social classification, norms and regulations are reproduced through the body. So, for example, childbearing and menstrual bodies are explained to be 'dangerous' (and hedged by taboos) because the flow of reproductive fluids represents a transgression of bodily boundaries. And women who are in these (liminal) states are themselves subject to social regulation. Gender hierarchy is thus explained through a much wider association of bodily substance to societal classifications of dirt/ pollution. More recently, work on gender hierarchy and body substance emerges in the work on the ageing body in India (Lamb 2000). Lamb discusses gender hierarchy in terms of the bodily metaphors of hot/cold and wet/dry in Northern India. Here women's procreative bodies tend to be associated with heat and wetness when they are 'open' to the passage of reproductive fluids. The 'closing off' of the body comes with age and the cessation of menstruation and is associated with becoming 'pure'. Women are seen to gain in social status once their bodies have been through reproduction, when over time they become 'cold' and 'dry', like the bodies of men.

Carol Counihan (1999) provides us with another comparative view of the body as gendered and sexed. She bases her analysis on the kinds of substances that flow between bodies and in terms of 'body permeability' and social power. Counihan uses the idea of permeable or the 'open' versus non-permeable or the 'closed' body as a way of understanding the relationship between women and men to be either equal or hierarchic. So, for example, she suggests that gender equality exists where men and women's bodies are reciprocally permeable enabling food and sexual substance to mutually flow between them. In the 'tribal' societies of New Guinea and the Amazon during sex, women not only receive semen but also penetrate men with a (threatening) female essence. This cross-penetration reduces gender hierarchy as it facilitates a circular flow of substance: women give men food to make semen which men return to women to make the fetus (also noted by Keesing and Strathern 1987). By comparison, Counihan suggests that, in the U.S.A., men and women's bodies are not mutually permeable. The fact that women's bodies are permeable while men's bodies are not, contributes to the hierarchical relationship between them. In general, Counihan suggests that the closing off of the body through sexual, reproductive or feeding practices is connected to a striving for power and identity. For Counihan the denial of penetrability in relation to food and sex (the closing off the body) is most powerfully evident in the case of anorectics.

In other contexts of gender differentiation, the body may be perceived as 'closed off' through an excess consumption, as opposed to a denial, of food. Poppenhoe's recent study of female fatness among the Azawagh Arab, pastoral nomads in the Sahara desert links the high value placed upon fatness to its central role in defining women as sexually desirable within Islamic societies (2004).[3] Azawagh girls are fattened on milk and porridge for a number of years before puberty to attain a figure that is admired by men and women in their society. Women work to maintain their fatness through the constant consumption of energy-dense foods. Azawagh women's bodies, Poppenhoe suggests, are both permeable and closed. They are permeable in that their bodies transform men's milk (from the animals they own) into making their own bodies sexually desirable. After their marriage this fat/wealth is passed on to their children in the form of nourishment. Women's fatness is not only valued by men but also highly regarded amongst the women themselves as it 'swells' the body shut, enabling them to get closer to the Islamic ideal of the whole, contained and therefore pure body.

Unlike Counihan, Poppenhoe suggests that desire created through the fattening of Azawagh women is associated with a range of gendered social dynamics not limited to sexuality alone. In other words sexuality is but one idiom through which fatness is conceived as desirable. This theme is further explored in the chapters in this volume by Walenkowitz and Randall who have worked among groups of Tuareg pastoralists similar to those described by Poppenhoe. The positive valuation of women's fatness as connected only to male sexual desire is an issue which is questioned by these authors who pay close attention to the gendered experiences of the fat women they write about. Here, it is clear that the gender roles enabled through being fat (e.g., in the ability of fat women to provide hospitality: Walenkowitz), were considered more desirable by women, compared to the hard work of thinner women such as the Bella slaves (Randall's account in this volume). The inequity between women highlights wider class differences operative in pastoral societies. As Randall points out, Tuareg women can sustain their fatness only because they are in an economic position to have Bella slaves to carry out their daily domestic chores, including the preparation and feeding of their women employers. Here fatness acts in establishing class distinction (in Bourdieu's sense, 1984).

In terms of gender distinctions, by contrast, Walenkowitz makes the important point in her chapter that a gendered focus on fatness overall, rather than on the force-feeding of women or in terms of male sexual desire alone, enables us to see the more equitable nature of relations between men and women. The Tuareg women she studied enjoyed both economic and reproductive autonomy through their fatness and through a kinship focussed on sibling ties. Through their generosity and hospitality made possible through an access to the wealth (mainly in animals) of their brothers, women were able to facilitate wide social and political networks. Here, we find that large bodies, as Gremillion suggests, 'signal a proper, ongoing flow of goods and services' that focus on shared social rather than individual gain (2005: 17). Fat women also command respect: Tuareg women made decisions about the place of birth and birth spacing. As Walenkowitz suggests, their immobility was not seen as a weakness, but more in terms of their power to induce mobility in others. Women's experience of fatness as a condition of sexual and reproductive autonomy is then critical to the shaping of their individual as well as collective sense of self.

Societal values attached to body size differ in relation to the different contexts of production and according to the abundance of, or conversely, the shortage of food experienced by the communities who live in these settings. Ideas of fatness prevail even in contexts of food shortage as Mabilia discusses in her chapter. It is in the contexts of everyday malnourishment that more subtle differences are made between what it means to be thin or fat. Mabilia describes how the Gogo distinguish between good and bad thinness, where to be 'too thin' not only indicates a lack of food but even more importantly, a poor incorporation into networks of social exchange. To be too fat ('bad fat') is when babies are large but 'weightless' which occur in conditions of extreme malnourishment and clinically characterised as Kwashiorkor, for example.

In other contexts of malnutrition such as in rural Rajasthan, north-west India, where I have worked, similarly nuanced differences arise in relation to body size. In this context it is widely believed by women that 'fat mothers make thin babies' (Unnithan-Kumar 2002). They not only bear smaller children but also are unable to nourish them well after birth as the quality of their breast milk is poor (described as *katna*, literally reducing). A distinction is made between *katna* breast milk and the *phulna* (swelling) milk produced by thin women. Fatness is regarded, above all, as a sign of laziness. In a poor economy such as Rajasthan where there is a high dependence on women's work, it is not difficult to understand why fat mothers are stigmatised (rather than valued as symbols of wealth and standing as among the Tuareg).

The societal value placed on maternal body size is thus variable according to the role women play in the production system more generally. As Martin (1987) and Bordo (1993) so clearly remind us, modern post-industrial capitalism is predicated upon the leanness and fitness of the bodies of its male and female workers. Being fat in this context is threatening to the economic functioning of society. Pathological bodies (of bulimic and anorectic individuals) are produced through a discourse which places contradictory demands on the individual, who is simultaneously encouraged to be self-indulgent and to be self-disciplined (Scheper-Hughes and Lock 1987). For many of the poor in these post-industrial contexts, a slim

and healthy body is out of reach as nutritious food prices equate
with those of luxury goods.

Fatness in less prosperous industrial contexts is also becoming
associated with ill health and its medical regulation as 'obesity'.
What is the impact of the traffic in medical concepts and regulatory
frameworks across the globe on the different societal valuations of
maternal fatness? To consider this issue we need to first deliberate
upon the relationship between fatness, food and fertility in bio-
cultural and nutrition perspectives.

Fertility, fatness and food: the body as object and agent

The relationship between fatness and reproduction is of significant
research interest to nutritional anthropologists and reproductive
biologists. In such bio-cultural perspectives, female fatness is
regarded primarily as evidence of reproductive potential. Frisch
(2002), for instance, suggests that body fat regulates the organs
responsible for reproduction. Women need a certain 'threshold'
amount of body fat to become, and remain, fertile. A weight gain or
loss at variance from this threshold level is, according to Frisch,
enough for women to experience the onset or absence of their
menstrual cycles. Body fat determines menarche (the onset of
menstruation) with plumper girls on average experiencing earlier
menarche and related sexual development, compared to thinner
girls. The mechanism through which such bodily changes are
triggered is related to the ability of body fat to convert male to female
hormones. What is of interest here is the positive valuation of fatness
within biologically based accounts of reproduction, as compared to
the current clinical perspective of fatness as detrimental to fertility
and procreative capacity.

The physiological implications of too much fat, manifested for
example in conditions such as infertility, is less understood, according
to Frisch, primarily because such bodily conditions are more recent
in the history of the human species (2002: 7). There is varying
cultural recognition given to the role of fatness in causing infertility.
In her chapter on Tuareg women, Randall suggests that the concerted
efforts at the production of fat bodies from as early an age as eight
years is not regarded by the Tuareg as affecting the ability of their
women to bear children. There was, however, recognition among
the Tuareg that fat girls were likely to reach puberty earlier, enabling
them to become sexually available to men sooner. As Randall notes,

female fatness was linked to sexual relations in the Tuareg groups she studied but it was not necessarily connected to shifts in the capacity to bear children.

The relationship between fat reserves in females and fecundity is, as Quandt observes, the most widely accepted biomedical indicator of the nutritionally thrifty adaptation of the human body. It is the strong selective pressures for energy efficiency and storage at times of food shortage, a constantly recurring feature in human history, that explain the operation of bodily mechanisms that today produce obesity (1996: 285). The role of maternal nutrition is critical not only in maintaining fertility but also for the health status of the mothers and the baby during pregnancy, and as an energy source for lactation. According to Quandt, there is a continuous circular relationship between mothers and daughters such that the nutritional status can be perpetuated over generations (such observations also emerge in recent biomedical research on maternal obesity as Heslehurst discusses in her chapter on cyclical obesity in this volume).

In general, cultural practices to do with food have been regarded in mainstream nutrition perspectives as opposed to the promotion of appropriate child health or reproductive health (Quandt).[4] De-Graft Aikins (Chapter 7, this volume) explores this disjunction as it emerges between healthcare providers and Ghanaian women over-nutritional intervention during pregnancy. Ghana is a country where there is a co-existence of significant levels of food insecurity and malnutrition among the poor while its wealthier sections face problems related to over-nutrition. Focusing on what she terms the 'social nutrition logic' operating amongst families and at a community level, de-Graft Aikins examines the everyday food practices, and what a cross-section of women 'think, feel and do about food' during their pregnancies in different regions of the country. She shows how this social logic is very differently configured compared with the 'medical nutrition logic' which underlies healthcare interventions and which characterises lay nutritional knowledge as 'incorrect'. Paying close attention to the kinds of foods pregnant women consume and the explanations that accompany their consumption, de-Graft Aikins suggests that the dichotomy between 'correct' and 'incorrect' community based nutrition knowledge is, in fact, too simplistic. This is because people tend to integrate ideas from the medical and social worlds, and across bio-medically framed and customary nutritional logics, depending on their geographical location, economic standing, education, historical context and inter-generational ties. Understanding how the social nutrition logic operates is critical to understanding why there is a significant degree of non-compliance to nutritional health interventions in the antenatal period amongst rural and urban poor women.

An ethnographic focus on food-related behaviour is critical in identifying differences between what people say they might do and what they actually do in food consumption. There is a wide range of responses to indigenous as well as biomedical nutritional advice given to pregnant women as several contributions in this volume show. The diversity in responses could, for example, be related to how individual women react to the unfamiliar taste of the medicine. A common reaction of pregnant women to iron folic acid tablets distributed by the auxiliary nurse midwives in the area of Rajasthan where I worked was a disgust at the taste and a consequent low uptake of this antenatal prescription. The same could also apply to indigenous preparations. Amongst the Hindu caste and Muslim women I worked with, it was mandatory for post-partum pregnant women to ingest a special sweet meat (*ladoo*) said to contain a mix of nutritive ingredients. The women often said they did not like the taste and therefore did not consume the *ladoo* (which were eaten by their children instead, Unnithan-Kumar 2002).

A cultural and subjective focus on food consumption brings new insights in its regard for the body as an agent in constructing and addressing what is the best form of nutritional intervention. In her chapter, de-Graft Aikins, for example, describes in the form of Ghanaian women's explanations for the food they consume as determined by the changes that occur in their bodies: whereby sensitivity to smell or a feeling of nausea leads them to crave or reject certain kinds of foods. An embodied perspective on nutrition is of central concern to the chapter by Clarke, who discusses the responses to an obesity related, nutritional health promotion programme initiated by the Irish Government amongst a poor community living in south-west Dublin. Clarke focuses on a community-based nutrition project, the HFME (Healthy Food Made Easy), as a means to examine the influence of the programme on the daily food practices of individual female participants. She finds that resistance to the programme is primarily based around issues of authority and ownership related to the knowledge participants have of their bodies' nutritional requirements. For one respondent, Brenda, it is her body which 'tells' her what to eat. Configuring her body as a source of authoritative knowledge on nutrition enables Brenda to contest the authority of the nutritional health practitioners running the workshop.

Forms of embodied knowledge, because they drive local responses to nutritional intervention, need to be recognised in nutritional programmes for antenatal care. In the following section, we consider further the competition or convergence of different forms of authoritative knowledge in the context of a globalising discourse on fatness.

Globalising fatness and authoritative knowledge: science, risk and health

State, scientific and medical perspectives and interventions to do with obesity are increasingly authoritative and yet in continual flux. Along with the media and fashion industry, they powerfully influence the ways in which maternal body size is perceived and experienced. Occurring globally such processes contribute to the standardisation of fatness as a pathological consumption of unhealthy foods to be redressed through a uniform set of regulatory practices, ideas, techniques and interventions around health and diet.

Obesity as understood in nutrition and health policy literature is linked to the modern lifestyle of people in the post-industrial capitalist world. Many countries in the developing world are believed to be undergoing similar economic and lifestyle transformations as reflected in the term 'nutrition transition'. The transition is signalled through the shift away from a diet high in fibre and common carbohydrates to a more energy-dense diet. Poorer industrialising countries such as India and Ghana, as discussed in this volume, demonstrate conditions which favour both an increase in obesity as well as under-nutrition (chapters here by de-Graft Aikins, Sridhar, Guntupalli). It is in urban localities and amongst the middle class where the dietary patterns are changing the fastest, as Sridhar describes. What is most striking about this transition is the change in the perceptions of what constitutes a healthy diet in these contexts. Energy, fat-dense and sugary foods, considered luxuries in the past when less easily available, have come to be seen as symbols of modern consumption and lifestyles. Examining the critical role of multinational food and drink companies such as McDonald's and Coca-Cola in this transition, Sridhar observes how they not only thrive on the new found association between fat and modernity, but contribute to it as well. The ability of companies such as McDonald's to reinvent themselves for the Indian market through what Sridhar calls 'hybrid products' is epitomised in its production of the Maharaja Mac: the burger without beef.

Discussions of the nutrition transition in the developing world also fuel a concern among the public as well as in policy circles, with health-related problems that accompany such changes in consumption. There is an increasingly shared language of the 'risks to health' amongst medical doctors, nutritionists and the state across the post-industrial and industrialising countries wherein scientific and medical observations of overweight are negatively correlated with health and fitness. The production of new categories of patients at risk from being overweight reflects particular social processes which for Gilman can be traced back to their origins in the mid-nineteenth century in the West (2008). They are, as Beck and Giddens suggest, social processes reflective of the conditions of late or high modernity (Beck 1986/1992; Giddens 1991; Caplan 2000).

Theorisations of risk as put forward by Giddens and Beck are useful in thinking about the concerns and counter concerns that have arisen around fatness as a health issue. There are two issues in particular which are of relevance here. First, the destabilising nature of risk discourse in the modern world where scientific certainty is both accepted yet increasingly contested. Second, the individual nature of risk response, by which people are forced to negotiate their life-style choices (Caplan 2000). Let us examine these two points in relation to health-related information about obesity produced, for example, in the U.K. In their recent document on *Tackling Obesity in England* (National Audit Office 2001), the British Government has expressed its worry over the trebling of obesity within the general population between 1980 and 2001. In the document, the standard measure used in calculating such prevalence is the Body Mass Index (BMI) in which the weight of a person in kilogrammes is divided by the square of their height in metres. A BMI of 20 to 25 is considered desirable with anything above this measure being regarded as overweight, obese and morbidly obese according to a person's increasing weight. Other categories of persons considered to be at risk are: children of obese parents, school children, poorer individuals especially women, and Black, Caribbean and Pakistani women. According to the NAO document these new, at risk, bodies are to be managed primarily through state programmes to do with diet, and exercise implemented within schools and communities which are considered especially vulnerable (discussed by Heslehurst, this volume).

The incidence of maternal obesity is the new subject within this climate of anxiety and intervention. Public-health data on the rates and trends of maternal obesity in the U.K. and internationally is as yet emerging nevertheless it is already understood that more than half of all maternal deaths in the years 2003 to 2005 were among overweight and obese women (see Heslehurst). Pregnant women who are overweight are considered to be four times more likely compared to normal weight women to have gestational diabetes, be thrice as likely to miscarry and twice as likely to experience still birth (Channel 4 documentary, 2006). Obese women, media reports suggest, are likely to have a two to three times greater propensity to undergo caesarean births compared to women with a regular BMI level (2006). Furthermore, obese women are being singled out as having reduced fertility, and if they are participating in assisted reproductive programmes, are actively encouraged to lose weight to increase their chances of successful conception. Negative associations have also been made between alcohol and tobacco consumption and childbirth among obese mothers. Babies of obese mothers are more likely to be born with additional health risks to do with early birth, cerebral palsy, deafness, blindness. And finally, there is a predisposition for the children of obese parents, especially where mothers are obese, to become obese themselves (Heslehurst's discussion on the 'obesity cycle', this volume).

In her chapter, Heslehurst engages in a debate from within the health services on identifying the causes and related interventions to do with maternal and childhood obesity. In particular she examines the ways in which maternal and childhood obesity could be connected at the stage of fetal development in the pregnant mother who is obese. Drawing on her research among women in Middlesborough, she suggests that the current health focus on the exogenous (e.g., lifestyle) factors as linked to obesity, is limited. She argues instead for the importance of endogenous factors, i.e., what is happening in the pregnant woman's body *in utero* or even prior to conception, as determining childhood obesity. Initiating obesity interventions in childhood are thus not early enough to prevent the increasing prevalence of obesity. The focus on the intra-uterine environment as a context where both taste and smell are developed by the fetus gives us a more nuanced understanding of the role of nutrition in the intergenerational transmission of obesity, although, as Heslehurst observes, the exact mechanisms as to how endogenous factors proliferate the development of obesity in offspring is unknown. Heslehurst's work in this volume brings us up to date with some of the current scientific research that is occurring on maternal obesity. It does, however, make invisible the host of social and emotional factors which induce parents to overfeed their children (Throsby 2007).

Heslehurst's discussion also shows us how, as in other contexts of biologically based work, obesity research is moving to focus more and more 'into' the body, at the molecular and genetic levels. The notion that an individual may be genetically predisposed to an illness, or obesity, has implications for their social constitution as persons and for their relations (or biosociality; Rabinow 1996; Gibbon and Novas 2008; Rose 2007) based on this. The fact that bodies are 'fat prone' on the basis of their genes, however, does not completely absolve the individual from the moral responsibility of their fatness. It is instead, as Throsby observes, 'the failure to do something about their weight which becomes the site of moral closure' (2007: 1565).

Going back to the observations made earlier on in this section where late modernity generates society characterised by a culture of risk, let us examine how such messages to do with obesity are actually 'read' and experienced by the public. What emerges from social-science-based accounts, including those in the present volume, is that people are often well aware of what constitutes healthy eating and the ill effects of an excessive consumption of fatty foods, but do not necessarily follow the health-related instructions themselves (as the Ghanaian or Irish examples in Chapters 7 and 8 indicate). There is also the prevailing idea that the risks posed to the individual by their inappropriate consumption can be dealt with by the advances in science itself (with technology facilitating a control over the body). Recent television and newspaper coverage in the U.K. which documents the successes of the gastric band and stomach stapling procedures to drastically inhibit the food intake of those individuals with serious weight problems is one such example. Undertaking weight-loss surgery is another example (Throsby, 2007).

The underlying reasons for the overconsumption of unhealthy foods can range from, on the one hand, knowing that it is not always the case that fatty foods are bad for you (also see Caplan 1997, for example), to the idea that the health messages are themselves misconstrued, as Aphramor and Gingrich, two U.K.-based dieticians, argue in their chapter in this volume. Dietetic understandings of obesity, Aphramor and Gingrich suggest, are uncritically preoccupied with a reduction or normalisation of body size. Central to this focus are two kinds of knowledge-based certainties: the notion that fatness is a reliable indicator of poor health status, and the concept of the energy balance thesis which posits that consumption should be less than energy expenditure and that thinness is a condition linked to better health. These concepts are further extended to apply to notions of what ought to be an appropriate size for pregnant mothers and infants.

Applying a critical perspective to the 'anti-fat' and energy balance approach of current dietetics, Aphramor and Gingrich suggest the need for dieticians to understand the production, reproduction and circulation of knowledge about obesity as the first step in reconstructing a more meaningful dietetics. They call for the need to embed nutritional discourse in people's local worlds and within the wider matrix of knowledge, care and constraints that underlie people's nutritional behaviour and desires. In her chapter on poor Irish women's resistance to nutritional directives for healthy eating, Clarke also critiques the individualising nature of public health policy on obesity, which lays the responsibility and blame on the individual person with no consideration given to the wider environmental factors such as poverty and the absence of accessible food outlets which constrain people to eat poorly. It is imperative that these messages circulate as widely as do biomedical warnings relating to the negative effects of overconsumption.

Overall, the volume offers a cross-cultural and multidisciplinary insight on fatness as it is perceived and experienced from the vantage of the maternal body. In the last few lines of this introductory chapter, I provide a more chronological summary of the chapters. In Chapter 2, Warin, Moore and Davies discuss the language and representations used by Australian women to characterise their bodies which is at variance with the clinical categorisation of them as obese. Describing the ways in which women draw upon certain maternal representations of their bodies ('nature') to legitimise their body size, the authors show how ideas of nature are used in the everyday discourse of fatness. Next, in Chapter 3, Randall examines the connections made between fatness and fertility in demographic and biosocial perspectives through her long-term study of the Tuareg nomadic pastoralists. The fattening of women is highly desired but contingent upon the economic means to hire domestic labour, and thus predicated on unequal class relationships. It is a context where the fatness of women is socially valued and sexually desired. No explicit connections are made between fatness and fecundity, or infertility. Fatness among Tuareg women is also at the centre of analysis in Chapter 4 by Walentowitz. She argues that in order to understand the implications of fatness for Tuareg men and women, maternal body size has to be seen in relation to the production of social relations (such as those between siblings as well as between husband and wife). In Chapter 5,

Mabilia further elaborates upon the connection between physical and social reproduction to show how ideas of fatness work in contexts of malnutrition and food scarcity. Through a focus on the moralities that surround rural, Tanzanian mothers' sexual and breastfeeding practices she shows how the shape, size and quality of the baby is a product of the flows of substances (sexual and nurturing: milk) between the bodies of mother, father and child. The inter-generational transmission of fatness is a theme that also informs Chapter 6 by Heslehurst, although more from a biomedical than sociocultural perspective. Heslehurst draws on clinical research on maternal obesity to show how fat mothers reproduce fat children through endogenous (fetal conditions) rather than exogenous (lifestyle-related) bodily conditions.

Shifting towards nutritional perspectives, in Chapter 7, de-Graft Aikins raises the important issue of the different medical and social views on nutrition as they emerge in pregnant Ghanaian women's food consumption practices. Focusing on the preferences and taboos surrounding raw and cooked food, de-Graft Aikins suggests, provides an understanding of the disconnections that exist between medical nutritional interventions during pregnancy and women's lack of compliance with these. Clarke, in Chapter 8, examines the resistance to nutritional policies that stem from Irish women's understandings of their bodies, their classification of food and the role that food plays in constituting their physical well-being. In Chapter 9, Sridhar charts the nutrition transition as it is occurring in India both in terms of what people consume but also in relation to their perceptions of what constitutes healthy eating. She critically examines the role of transnational food corporations such as McDonald's in the globalisation of food consumption patterns. Guntupalli in Chapter 10 looks at the increasing reproductive health concerns of Indian women that emerge across the different regional, socio-economic and demographic categories. The volume ends with Chapter 11, with a robust, feminist challenge posed by Aphramor and Gingrich, two nutritional practitioners and activists, who question the validity of dietary restrictions in enhancing the well-being of those considered obese.

Notes

1. The well-known work of Foucault, for example, examines the institutional mechanisms through which power is exercised on individual bodies and how populations are managed by the state. Bourdieu (1977) has looked at more socially embedded institutional mechanisms which

shape body practices and perceptions. Mary Douglas's work (1979) has used the idea of the body as symbolic and representational of a particular society's norms and values and related cultural anxieties: anxieties manifest in the societal taboos regarding bodily substances, fluids, orifices and boundaries (Scheper-Hughes and Lock 1987; Csordas 1994).

2. A note on terminology: we use the term 'obesity' specifically when referring to medical understandings of the term. We use 'fatness' following Kulick and Meneley (2005) as a means to incorporate the diverse constructions of what fatness means in any social context, and in order to distinguish it from clinical terminology such as obesity (and the linear way in which the term is used to describe an individual in terms of a scale from normal to morbidly obese). Some writers (e.g., Monaghan 2008) consider fatness to be a term loaded with negative cultural meaning, preferring a more functionally ambiguous term such as 'big' or 'large' (see also Warin, Moore and Davis, this volume). Writing from the perspective of the U.K. men he interviewed, Monaghan uses the positive, self-perception of his respondents as 'big' to fit in better with their ideas of masculinity and respectability in the fat-phobic society in which they live.

3. There are other groups in the world (e.g., Egypt, Jamaica, Nigeria, India, New Guinea) that share the value placed on women's plumpness as mothers but as Poppenhoe suggests, in Moor society, fattening is a central preoccupation of women alongside childbearing and extends over a long period.

4. Nutritional discrimination of girls in favour of boys has been widely cited to occur in cultures where men are regarded as socially superior to women (Quandt 1997).

References

Beck, U. 1992 (1986). *Risk Society: Towards a New Modernity.* London: Sage.

Bordo, S. 2003 (1993). *Unbearable Weight: Feminism, Western Culture and the Body.* Berkeley: University of California Press.

Bourdieu, P. 1977. *Outline of a Theory of Practice.* Cambridge: Cambridge University Press.

—— 1984. *Distinction: A Social Critique of the Judgement of Taste.* Translated by R. Nice. Cambridge: Harvard University Press.

Caplan, P. 1997. *Food, Health and Identity.* London: Routledge.

—— 1999. *Risk Revisited.* London: Pluto.

Counihan, C. 1999. *The Anthropology of Food and Body: Gender, Meaning and Power.* London: Routledge.

Csordas, T. 1994. *Embodiment and Experience.* Cambridge: Cambridge University Press.

Douglas, M. 1979. *Purity and Danger: An Analysis of Concepts of Pollution and Taboo.* London: Routledge.

Foucault, M. 1981. *History of Sexuality*. Vol. 1. London: Peregrine.

Franklin, S. 2001. 'Biologization Revisited: Kinship Theory in the Context of the New Biologies', in S. Franklin and S. Mckinnon (eds), *Relative Values: Reconfiguring Kinship Studies*. Durham, NC: Duke University Press, pp. 302–29.

Frisch, R. 2002. *Female Fertility and the Body Fat Connection*. Chicago: University of Chicago Press.

Gard, M. and J. Wright. 2005. *The Obesity Epidemic: Science, Morality and Ideology*. London: Routledge.

Gibbon, S. and C. Novas. 2008. *Biosocialities, Genetics and the Social Sciences: Making Biologies and Identities*. London: Routledge.

Giddens, A. 1991. *Modernity and Self Identity: Self and Society in the Late Modern Age*. Cambridge: Polity.

Gilman, S. 2008. *Fat: A Cultural History of Obesity*. Cambridge: Polity.

Gremillion, H. 2005. 'The Cultural Politics of Body Size', *Annual Review of Anthropology* 34: 13–32.

Haraway, D. 1993. 'The Biopolitics of Post Modern Bodies: Determinations of Self in Immune System Discourse', in S. Lindenbaum and M. Lock (eds), *Knowledge, Power and Practice*. Berkeley: University of California Press, pp. 364–411.

Jordan, B. 1997. 'Authoritative Knowledge and its Construction', in R. Davis-Floyd and C. Sargeant (eds), *Childbirth and Authoritative Knowledge*. Berkeley: University of California Press. 55-80

Jutel, A. 2006. 'The Emergence of Overweight as a Disease Entity: Measuring up Normality', *Social Science and Medicine* 63: 2268–76.

Kulick, D. and A. Meneley. 2005. *Fat: The Anthropology of an Obsession*. New York: Penguin.

Lamb, S. 2000. *White Saris and Sweet Mangos: Ageing, Gender and Body in North India*. Berkeley: University of California Press.

MacCormack, C. and M. Strathern. 1980. *Nature, Culture and Gender*. Cambridge: Cambridge University Press.

Martin, E. 1987. *The Woman in the Body*. Milton Keynes: Open University Press.

——— 1994. *Flexible Bodies: The Work of Immunity in American Culture from the Days of Polio to the Age of Aids*. Boston: Beacon Press.

Monaghan, L. 2008. *Men and the War on Obesity: A Sociological Study*. London: Routledge.

Ortner, S. 1974. 'Female is to Male as Nature is to Culture', in M. Rosaldo and L. Lamphere (eds), *Women, Culture and Society*. Stanford: Stanford University Press, pp. 56–74.

Poppenhoe, R. 2004. *Feeding Desire: Fatness, Beauty and Sexuality among a Saharan People*. London: Routledge.

——— 2005. 'Ideal', in D. Kulick and A. Meneley (eds), *Fat: The Anthropology of an Obsession*. New York: Penguin, pp. 9–29.

Quandt, S. 1997. 'Nutrition in Medical Anthropology', in C. Sargeant and T. Johnson (eds), *Medical Anthropology: Contemporary Theory and Method*. Westport, CT: Praeger, pp. 272–93.

Rabinow, P. 1996. *Essays on the Anthropology of Reason*. Princeton, NJ: Princeton University Press.

Rose, N. 2007. *The Politics of Life Itself: Biomedicine, Power and Subjectivity in the Twenty-first Century*. Princeton: Princeton University Press.

Scheper-Hughes, N. and M. Lock. 1987. 'Mindful Body: A Prolegomenon to Future Work in Medical Anthropology', *Medical Anthropology Quarterly* 1: 6–14.

Sontag, S. 1989. *Aids and Its Metaphors*. New York: Farrar, Strauss and Giroux.

Strathern , M. (ed.). 1987. *Dealing with Inequality: Analysing Gender Relations in Melanesia and Beyond*. Cambridge: Cambridge University Press.

———— 1992. *Reproducing the Future: Anthropology, Kinship and the New Reproductive Technologies*. Manchester: Manchester University Press.

Throsby, K. 2007. 'How Could You Let Yourself Get Like That?: Stories of the Origins of Obesity in Accounts of Weight-loss Surgery', *Social Science and Medicine* 65: 1567–71.

Unnithan-Kumar, M. 2002. 'Midwives among Others: Knowledges of Healing and the Politics of Emotions in Rajasthan, Northwest India', in S. Rozario and G. Samuel (eds), *Daughters of Hariti: Childbirth and Female Healers in Southand Southeast Asia*. London: Routledge, pp. 109–30.

Chapter 2

THE TRAFFIC IN 'NATURE':
MATERNAL BODIES AND OBESITY

Megan Warin, Vivienne Moore and Michael Davies

Introduction

In a recent article on the 'obesity debate', Lee Monaghan sets up the key dimensions of a problem that this chapter aims to address. As a sociologist who has written extensively about masculinity, weight-related issues and embodiment, Monaghan was invited by editors of the *Men's Health Forum* ('a magazine distributed to health professionals, politicians and policymakers in England and Wales' (2005a: 302)) to write a counterpiece to a clinical discussion on why he thought the case against obesity was overstated. The positioning of opposing arguments is a favourite journalistic device, and one which Monaghan, justifiably, felt uneasy with. He was being asked to reject the highly publicised obesity epidemic and, in the same vein as Campos's (2004) popular work, critique the taken-for-granted alarmist assumption that obesity is a 'massive public health problem that should be tackled' (Monaghan 2005a: 303). Despite Monaghan's assertions that he was not taking sides or claiming that 'being fat' is beneficial to health, editorial changes to the title of his article: 'Taking sides: can fat be good for you? Yes' (2005b), reduced the argument to an essentialist either/or position.

We start with Monaghan's example as it throws light on the ways in which the complexity of obesity (and all its differing and competing claims) is simplified. Since the discursive inception of the 'obesity epidemic' in the late 1990s there has been an exponential growth in writing about obesity, fatness and bodies. The main proponents in this field include clinicians, public health specialists (including epidemiological and critical approaches to health promotion), psychologists, nutritionists, feminists, sociologists, and to a lesser extent, anthropologists and social geographers. Despite these multiple perspectives (and the acknowledgment of the complexity of factors that contribute to the onset and progression of 'obesity'), Gremillion makes the important point in arguing that 'scholarly interest in the shaping and signification of bodily forms is linked to culturally and historically specific preoccupations regarding the production, assessment, and management of bodies' (2005: 26). These preoccupations, we suggest, have been reduced to the 'quarrelsome science wars between those working in the borderlands between the natural and social sciences' (Labinger and Collins 2001: ix, cited in Monaghan 2005a: 303). In focusing on obese women's experiences of corporeality, this paper examines the traffic in concepts of nature/culture as they are used in everyday contexts, thus challenging a taken-for-granted and organizing epistemological divide that dominates the current literature.

We support Saguy and Riley (2005: 869) in their claim that 'the so-called obesity epidemic remains a highly contested scientific and social fact'. While mainstream medical literature acknowledges multiple causes for obesity, including genetic, behavioural and 'cultural influences', it is the behavioural variables that are strongly favoured. Drawing upon a wealth of clinical research, the *Surgeon's General's Call* echoes the underlying principle of the energy balance equation: 'For the vast majority of individuals, overweight and obesity result from excessive calorie consumption and/or inadequate physical activity' (Kwan 2006: 11). Appeals to action and intervention are thus framed by individual changes to 'lifestyle'.

In critiquing clinical and epidemiological evidence, social constructionist approaches have questioned the ways in which fat is presented as fact (for example, see Gard and Wright 2001, 2005; Campos 2004; Campos et al. 2006; Aphramor 2005; Rich and Evans 2005) and pointed to the underlying ideological, political and moral agendas of such evidence. It is tempting to simply characterise this group as trained in the humanities or social sciences and thus inclined to critique and question the foundations, claims and discourses of 'scientific' and biomedical knowledge, yet this group is drawn from a broad array of disciplines (including those trained in medical and health sciences).

While epistemological differences divide and drive the 'obesity wars' and provide a foil for counter argument (Neumark-Sztainer 1999), there remains a reluctance to problematise or interrogate the premise of these differences. Saguy and Almeling (2008), for example, in their examination of how 'overweight' and 'obesity' are being defined by claims-makers as social problems (specifically, the media reporting of medical science), do not seek to intervene in the very debates that premise their argument. Similarly, Monaghan, in his study on male bodies and obesity in the north east of England, acknowledges the conflicts that have arisen in the challenges to orthodox views, yet states that he doesn't want to engage with the 'institutionalised war on fat' (2007: 584). His focus is on the gap between the authoritatively defined claims of obesity, and men's critical understandings of embodied masculine identities. In presenting and honouring 'men's justificatory accounts for levels of body mass that medicine labels too heavy, Monaghan's aim is to unsettle 'mainstream obesity literature and health promotion' (2007: 606). The authors of this current chapter too, have previously distanced themselves from the 'epistemological debates of biological fact versus social constructionism in obesity discourses' (Warin et al. 2008: 2).

This chapter questions the very analytical approaches and a priori assumptions that underpin epistemological approaches to obesity. In a sense we 'work backwards' (Strathern 1999: 6) to problematise the organizing principles of the separation of the social from the biological (Latour 1993). This separation, as Roepstorff, Bubandt and Kull (2003) similarly note in their ecological work on 'imagining nature', is directly engaged with a 'certain conceptual division of the world that relegates phenomena into distinct categories: nature on one side, society on the other' (Roepstorff et al. 2003: 19). Thus, we have debates over the biological 'grounding' of obesity as a pathology, disease or gene located in individual bodies, and a social constructionist focus on 'bodies' and their relationships to wider structural issues such as of poverty, gender and class. Even in analyses which attempt to bridge these dichotomies of individual and society (as in 'obesogenic' approaches), 'nature' (genetic propensity) and 'culture' (vaguely interpreted as environment or 'ethnicity') remain 'black boxed'.

The dichotomy between nature/culture has been a backdrop to anthropological and feminist theories for decades, and for social anthropologists, it is a truism to state that nature is culturally

Obesity: closer to 'nature'

The women in this study acknowledged that they were (and had been at varying points in their lives) 'big' or overweight, but not one of them, however, identified as obese (and the direct association with diseased, sick, and 'at risk' bodies). Andrea casually inserted this comment into our conversation one day: 'By the way, I am overweight, I know that. I just do not like to use the word obese.' We have previously suggested that this resistance to clinical categorisation is not surprising when you consider the profound stigma that is associated with bodies that are labelled as 'deviant' and 'out of bounds' (Warin et al. 2008). Goffman (1968) mentions the stigmatised category of obesity in his early work on spoiled identities, and Sobal (2004) notes how this stigmatisation has continued in post-industrial societies. Common, taken-for-granted associations with obesity – such as the ways in which 'fat people are stereotypically constructed as undisciplined, self-indulgent, unhealthy, lazy, untrustworthy, unwilling and non-conforming' (Bell and Valentine 1997: 36) – no doubt play a significant role in these women's rejection of the label. Yet, as we outline below, there were other significant frameworks that were articulated in order to mark difference.

Diagnostic categories and classifications are often resisted and mobilised according to differing politics and identities.[3] Monaghan found that his male respondents from the slimming club dismissed medicalised notions of 'appropriate' weight-for-height recommendations in favour of more meaningful, embodied understandings of heaviness (2007: 605). These understandings included; gendered accommodations of heaviness, healthiness and physical fitness; looking and feeling ill at a putatively 'healthy' BMI; and the irrationality of standardisation (Monaghan 2007: 605).

When asked what 'obesity' meant to the participants in our study they were quick to draw upon cultural frameworks of difference through 'the Other'. The elucidation and documentation of cultural difference has been a longstanding concern of anthropology, and operates in a number of differing fields: for example, Said (1979) on Orientalism; Fabian (1983, 2002) on time; or Torgovnick (1990) in her discussion of primitivism. What draws discussions of the Other together, is a set of implicit assumptions about difference and distance, in which that which is not 'me' or 'normal' is defined as 'inferior, deviant, subordinate and subordinateable' (Torgovnick 1990: 21). In describing obesity, the women in this project drew upon constellations of otherness in two ways. The first was via anthropomorphic representations of obese people. Briony, for example, immediately thought of a man she knew whose nickname was 'animal' because he 'was so huge from pigging out all the time':

> I was using my Mum's car, I think it was a little *Pulsar* and he got in and honestly I thought the car was going to topple over ... he was so big that he couldn't sit on the seat properly ... that's big ... that's when I thought he was going to die, he could hardly walk. He'd go through McDonald's drive-through and he'd order, like five Big Macs, three lots of large chips and stuff like that and he had legs like elephant legs ... he couldn't even get into a shower ... I mean what do you do, go outside and hose yourself down or what?

Briony drew on a historically produced analytical distinction between animals and humans, in which animals are positioned as inferior to humans, and closer to nature. As many scholars have already noted (Mullin 1999; Ingold 1988; Roepstorff 2001), the ways in which people think of and regard animals as 'non-human persons', tells us much about human social relations. Her friend is described as 'an animal' because he shares certain qualities with elephants and pigs, notably, his enormous size and 'out-of-control' eating habits. Pigs not only wallow in mud (which is in itself constructed as a 'dirty' and an 'in-between' substance), but they are also, according to Douglas's classic text on the taxonomic classification of Jewish dietary law, anomalous. And because pigs do not fully conform to their class (in that they are cloven footed but do not chew the cud) they are forbidden and avoided (1966,1984: 55). In Briony's representation and in popular idioms and discourses, an obese body is a body that is conceptually out of bounds and transgressive (see also Evans and Le Besco 2001), and like a pig, uncivilised in its manners. Other participants made analogies between humans and animals to describe eating behaviours and bodies, such as 'scoffing it down like a starved dingo', to gendered accounts of women 'eating like birds'.

The second othering frame that participants drew upon was one of obesity as a 'freak show'; 'the public showing of misfits and wonders of nature' (van Dijck 2002). Not dissimilar to animals being exhibited in a zoo, obese people are often paraded on television shows and in popular discourses as medical spectacles (see also Thomson and Garland 1996). Michelle, a single parent of three young children, described her understandings of obese as constructed through popular discourse:

I've seen them [obese people] on the Jerry Springer show [they are] like three hundred pounds or whatever; they cannot get out of bed and they cannot do anything. That's big, that's huge. That's when you think for their health sake they've got to do something ... they had to cut out this guy's wall and get a crane to get him out and onto the Jerry Springer stage because he'd been stuck in that house the whole time and he couldn't get through the door or nothing.

Others similarly described obesity as 'the other side of fat', or 'when you need a forklift to come and carry you out of your house, that's obese ... that's *beyond* obese I think' (participant's emphasis). This 'other side' is a line, not marked by clinical distinctions between overweight, obesity and morbid obesity, but rather marks the crossing of an imaginary line whereby people do not conform to bodily and moral boundaries of personhood.

These representations of obesity as visual extremes assist in constructing the bodies of others as abject, excessive, unhealthy and diseased. Van Dijck argues that these images are contemporary versions of nineteenth-century freak shows, transformed via media into medicalised marvels, where the horror and fascination with bodies that are transgressive (as 'wonders or freaks of nature') are brought back into the realm of 'normality' (often through surgery) (van Dijck 2002).

In rejecting bodies that are socially constructed as Other (through both anthropomorphism and spectacle) these women were positioning their own identities. Representations of the Other are not one sided, but simultaneously describe the ways in which 'the self' is imagined. Othering, as Lucas and Barrett argue, is 'an essentially reflexive tool by which the West comes to understand itself' (1995: 313). In distancing themselves from discourses of obesity, the women in this project were creating a space in which their own understandings of weight and body shape could be positioned. As Maddy so clearly stated, 'obese is sloppy, slovenly and not me'. If obese is 'not me', how did these women understand their bodies?

Made by nature: 'big bones'

When simply asked how they would describe their bodies, all of the women used euphemisms. These included being 'big boned', 'large boned', 'heavy boned', 'short for my weight', 'chubby', 'cuddly', 'well built', 'stocky' and 'heavy'. Bones were frequently cited as a reason for heavy bodies: 'it's the bones you know, they're too big, I'm too big boned'. Rashida explained: 'I'm very big boned. I've always been big boned. I take after my father's side on that.' When I asked Donna her current weight in the context of her weight history and the upcoming gastric surgery, she lowered her voice and whispered:

> One-hundred-and-fifteen [kilos]. The doctors told me that my ideal weight is about eighty-two kilos, because I've got large bones ...
> [What do you mean by 'large bones'?]
> Okay, big bones: feel my wrist [she offers her wrist for me to feel the hardness and largeness of her bones]. Bones, not fat, bones ... you see, I've got bones. I can feel my rib bones.

Donna knew that her 'large bones' were the source of her 'heaviness', as she had been told by a 'health expert' as a teenager:

> I went and had X-rays when I was thirteen years old and she [the radiographer] noticed that my fingers are like big bones and she said to me 'that's the reason why you are sort of large for thirteen' ... that's the reason I'm so big and I'm thinking, 'no worries' ... most of my family, my Mum, my Nan, my little sister ... all my Pop's side, most of us are all big. I do not think there's one skinny sheila [woman] apart from people that are married into the family ... it's just a family tradition.

For Donna, the power of science (and in this case technology) inspired the logic behind her size (and that of her family).

Monaghan similarly found that the men he interviewed at the slimming club in England invoked the 'heavy bones and bone density argument' to rationalise their 'natural build' and critique the narrowness of the BMI (2005: 602). He suggests that these 'non-negotiable characteristics like 'body structure' or 'build' (Monaghan 2005: 601) are thus a gendered discourse of resistance to clinical parameters for they express men's justifiable dissatisfaction with the authoritative view that there are specific measures that they could or should conform to' (Monaghan: 602).

Why are bones 'non-negotiable'? We suggest that this reflects biology-in/as-culture at play. Interestingly, it is bones rather than flesh that are repeatedly cited as the reason for heavy bodies. While there is an extensive literature on the meanings of bones and flesh in Melanesian societies (Stewart and Strathern 2001), there is no comparable ethnographic work on the ways in which culture shapes bones in Western societies.[4] It is reasonable to suggest that bones (rather than flesh) are drawn upon to explain body weight as they are the unseen weight that does not carry the stigma or 'sins' of flesh. Flesh, particularly, on a fat person, is negatively constructed as sagging, wobbly, and wrinkly, in contrast to the hardness of bones. Bones are the body stripped bare of flesh and point to a deep, archaeological site of forensic evidence and history. Unlike the 'flabbiness' and corruption of flesh and fat; bones speak of 'truth' and strength. Big bones thus become couched in ways that mirror discursive constructions of new genetic knowledge and in particular, 'the notion that one's identity is an inborn, natural, and unalterable quality' (Brodwin 2002). In doing so, they distance the moral failings that are associated with popular representations of obesity.

This is not simply biological rationalisation, as participants couched the 'naturalness' of bones also in terms of social connections. Big bones are a shared heritage; they are 'family tradition'. In describing their own bodies women made explicit links to other large and thin bodies of family members (and who got 'the bad [fat] genes'). These links were often with mothers or women who featured as primary caregivers and the day to day mundane activities that they remembered about commensality. When mapping out her family tree, Emily, a young single mother who had been on 'hundreds of diets', described the 'people on my Mum's side of the family as big boned ... more so the females ... especially Mum and Nanna'. At the end of our interview Emily showed me a photograph of her brother and father sitting at her kitchen table for her daughter's recent birthday party: both who would be clinically described as morbidly obese. When asked if there might be a reason why members of her family might be big, she responded by saying that 'being big is part of the family', and for the family, 'this was nothing out of the ordinary'. Emily draws on an intersecting discourse of biology and relatedness to explain the inevitability of her own body size.

Participants also drew on traditional biological kinship explanations for their body shape, saying their 'shape', or 'build' (not bones), was 'inherited', 'hereditary' and 'genetic'. Alison, for example, spoke at length about her biological mother who had 'put her up' for adoption at birth. Her mother was a student nurse in Adelaide, Australia, and as an unwed, single woman in the 1960s she hid the pregnancy until she went into labour (coincidentally whilst at work in a maternity hospital). When Alison became a mother herself, she tracked her birth mother down (prompted by a desire to find out about an eye condition that she and her daughter have and any other possible genetic conditions). Alison laughed in remembering her first meeting with her mother: 'she was large ... she has the same build as me ... this is what my body will look like in twenty years time'. For Alison, the reunion allowed her to feel reassured that her own troubled relationship with her body was 'not altogether my doing' (in that she was 'built like [her] birth mother') and 'nothing is going to change this'. In effect, Alison was reinforcing traditional biomedical explanations of kinship that are anchored in genetic ties, flowing from past to future, possessing a permanence that transcends time (Finkler 2000: 43).

In locating nature in the intimacy of bones, genetic relationships and family traditions, Alison, Donna and Emily are describing a particular 'traffic in nature', in which a range of possibilities for inclusion and exclusion of personhood are possible. The traffic is not only located in what Crossley (2004) refers to as the 'biochemical concept of the person', but flows through kinship (both biological and social) lines and practices, and is expressed as an idiom of relatedness. Franklin makes a similar argument in her analysis of scientific facts in women's experiences of IVF (*in vitro* fertilisation). She argues that, in the West, 'facts' of biology symbolise two things: the possession of a particular form of knowledge, which offers a particular access to truth, and certain types of relatedness (1997: 208).

'Childbearing hips'

The most obvious type of relatedness for the women in this project was motherhood. Despite the extensive feminist and anthropological literature that seeks to problematise the connection between gender inequalities, women's bodies and nature (McCormack and Strathern 1980; Collier and Rosaldo 1981; Strathern 1988), the mothers in our project drew explicitly on their reproductive and nurturing capacities to explain their bodies. All but one of the women, 'loved being pregnant'. Jane, who was having her second child, described pregnancy as an excuse: 'now I'm pregnant I'm relaxed, I've got an excuse to be big'. This 'excuse' was related to an entirely different sense of embodiment.

Pregnant bodies and fat bodies are both subject to public scrutiny, but with entirely different consequences. One of the reasons the women in this project 'loved being pregnant', was that their big bodies (identified by others as obese), were replaced by a 'thriving, glowing and healthy body', which, as Amanda commented, 'was meant to eat, allowed to eat'. Upton and Han (2003) similarly describe this liberation from scrutiny in their analysis of pregnancy in American 'culture and work'. For women that are already large or gain significant weight during pregnancy, the pregnant body takes on completely different meanings around fatness and bodily transformation, in which, as Tamara said 'it's natural to put on weight when you are pregnant'. Others commented that it was 'natural to feel differently about your weight at different times, and particularly when you are pregnant'. Coming into motherhood provided an alternative framing of gendered bodies, in which 'childbearing hips' and 'big shoulders' (carrying emotional loads of maternal labour), signalled a capacity for reproduction and nurturing.

McBride (1989: 95) argues that constructions of mothering and motherhood are associated with broad shoulders, and the capacity to be competent, strong, generous and solid. In order to 'carry the weight of the world' McBride explicitly draws upon her own gendered and working-class values of nurturing, in which being large/kind-hearted, bounteous and hospitable are valued. This is in striking opposition to the 'spindly and shallow' ambitions of thinness that she sees as attributed to the embodiment of class mobility (McBride 1989: 95). The direct association of nurturing with largeness was a feature of working-class mothers in our study, and particularly for single working-class parents who described a need to be physically strong and big to protect their children.[5]

In their pride and positive evaluations of pregnancy and mothering roles, these women echoed early feminist views put forward by Chodorow (1978), Dinnerstein (1976) and Ruddick (1980), who championed the natural, biological processes of becoming a mother. Even with Strathern's (1992) incisive challenge of this taken-for-granted 'natural fact' and the more recent literature on NRT, dominant ideologies surrounding maternity in many countries continue to focus on the 'natural' role of women and mothers (Lewin 1995; Kahn 2000; Pashigian 2002; Teman 2003). Heavily pregnant Tamara stated that 'this is just the way my body is, and if I'm going to lose weight I'm going to do it naturally'. When I asked what 'naturally' meant, she replied 'when I'm not making babies'.

In the differing discourses of bones and kinship, nature was used to explain bodies that were larger due to biological and nurturing roles. This 'closeness to nature' was unlike obesity in Others (which drew on negative, moral associations of individual self worth), and couched in terms of inevitable genetic endowment or positive reference to the power of women's bodies to reproduce. The explanations for this intimacy provided a reason (it's my 'bad genes' or 'blame it on the baby'), thus distancing moral failings or individual culpability.

Conclusion

Our aim in this chapter has been to highlight the complex ways in which a group of mothers who are all clinically obese, distance themselves from this category. 'Nature' is a powerful organizing principle through which they locate obesity in the bodies of Others, in order to distance themselves from a personhood constructed as morally disgusting and diseased. In explaining their own body size (as big and overweight but not obese), nature is again called upon (in a different form) to explain the inevitability of genetic material, of 'big bones' and 'childbearing hips'. Nature thus emerges in three different guises, first, as a negative, anthropomorphic, non-human behaviour or spectacle in which obesity is located in the bodies of Others. Here, obesity is not part of these women's identities. A differing concept of nature is, however, located within 'the self' and used to explain 'big bones' and the naturalness of shared, genetic substances. Like the distancing of the Other, genetic reasoning absolves blame and individual failing in regards to weight. And third, nature is drawn upon to explain the power of nurturing, of the capacity to bear and nurture children, and the naturalness of associated weight gain and changing body shapes. In this traffic, nature is not positioned as a dichotomy, but moves in and through bodies, distancing and connecting relations.

In examining the content of what people say and do in relation to motherhood and obesity, it is impossible to ignore or sidestep the ways in which multiple concepts of nature are enfolded into experiences and used to rationalise changing body shapes and identities. Nature should not be dismissed as a social construction, as it can no longer be simply located on one side of the nature/culture divide. As recent theorising in reproductive technology has demonstrated, the ways in which biology works in and as culture is useful when bringing a new analytic light to obesity.

The reconfiguring of nature in these women's experiences forces us to question the premises upon which we enter the so called 'obesity war' debate. Those who argue that obesity is a scientific fabrication or an ideological invention (see, for example, Coveney 2008; Campos et al. 2005; Gard and Wright 2005) do not question the foundations of their argument. Taking a social constructionist position they take for granted that 'facts' belong to nature, and 'constructs' (or 'fetishes' for Latour 1993) to culture, thus reproducing the nature/culture opposition. In doing so, they ignore the ways in which the materiality of the body is both real and completely constructed at the same time. As Eva Berglund argues, 'we can deconstruct nature analytically, but in practice it remains a key ontological foundation, indefinite as it may be, for operating both intellectually and morally in a world in which we find ourselves today' (Berglund, responding in Escobar 1999).

Moreover, there is little acknowledgment in the obesity literature of how gendered bodies change across the life-course, and the shifting identities of different embodied processes. In obesity surveillance programmess bodies are measured and quantified and become marked according to these fixed valuations. Longitudinal and ethnographic studies, which can 'track' changes across time and space, mean that bodies can be studied not as snapshots, but as 'bodies-in-process', thus acknowledging the shifting and ambigious experiences of embodiment. Thinking of bodies as moving and shifting in the traffic of cultural systems allows for different expressions of agency, and the power to change perceptions of where bodies are located. Bodies, and in particular, women's experiences of embodiment, are constantly changing and cyclical, but redirected towards a metaphorical (and masculine) template of linearity and stability.

On a more practical level, this chapter has highlighted the disparity between clinical parameters of obesity and women's own identification and experience of their bodies. This disparity is vital for health promotion initiatives for a number of reasons. Mothers are becoming the new 'moral panic' in obesity discourses, either blamed for overfeeding or misfeeding their children (as in the recent case of 'abuse' brought against a mother of an obese child in the north east of England), or in the increased surveillance of maternal obesity (Zivkovic et al. 2010). We argue that it is important to understand the traffic in nature amongst mothers (and mothers to be), for producing negative discourses around 'obese mothers' feeds directly into very powerful intersections of mothering, reproduction and relationships. For the women in this project, to target their nurturing roles is an attack on an important aspect of their identity and relationships. Similarly, to ask the women to change their lifestyles or to eat less and exercise more does not accord with their own understandings of weight, which locate 'being large' in the realm of nature and outside of cultural interventions. Unlike the fixed and constraining parameters of the BMI, nature (as it intersects with biological and sociological processes) makes more allowances for, and legitimates, bodily differences. Most importantly, how might public health initiatives address an audience that simply do not identify with the 'problem', or see themselves as diseased?

Notes

1. For an excellent rethinking of corporeality and obesity beyond binaries, see Murray's (2008) phenomenological work on 'fat' bodily being, intersubjectivity and 'living the flesh'.
2. The study was entitled: 'Food and Families Project'.
3. See, for example, the literature on barebacking and bug chasing amongst gay men, and the desire to 'gift' HIV status and identity (Gauthier and Forsyth 1999; Tomso 2004).
4. Fausto-Sterling (2005), in her life-course systems approach to the analysis of sex/gender is an exception. She argues for a biocultural approach that examines how sociocultural categories act on bone production. 'Culture' she argues, is thus an 'interchangeable partner in producing body systems commonly referred to as biology' (2005: 1516).
5. This rationalisation echoes Susie Orbach's often cited work (1978) in which she argues that fat women maintain their shape as armour against the sexual demands and desires of patriarchal society.

References

Aphramor, L. 2005. 'Is a Weight-centred Health Framework Salutogenic? Some Thoughts on Unhinging Certain Dietary Ideologies', *Social Theory and Health* 3: 315–40.

Bell, D. and J. Valentine (eds). 1997. *Consuming Geography: We Are Where We Eat*. London and New York: Routledge.

Brodwin, P. 2002. 'Genetics, Identity and the Anthropology of Essentialism', *Anthropological Quarterly* 75(2): 323–30.

Campos, P. 2004. *The Obesity Myth: Why Our Obsession with Weight Is Hazardous to Our Health*. London: Penguin.

———, A. Saguy, P. Ernsberger, E. Oliver and G. Gaesser. 2006. 'The Epidemiology of Overweight and Obesity: Public Health Crisis or Moral Panic?' *International Journal of Epidemiology* 35: 55–60.

Carryer, J. 1997. 'The Embodied Experience of Largeness', in M. de Ras and V. Grace (eds), *Bodily Boundaries, Sexualised Genders and Medical Discourses*. New Zealand: Dunmore Press, pp. 99–109.

was a well-established institution in pre-colonial times with most Tamasheq slaves originally captured in raids on villages and other communities living further south. Bella are black African and although they now all speak Tamasheq, they clearly have different genetic origins to the Berber Tuareg. Many Bella were liberated in the colonial period and after independence, although de facto ownership of slaves still continued at the time of the 1981 to 1982 surveys, with many Tuareg having resident Bella to do most domestic and herding work. The 1981 and 1982 surveys included both domestic servile Bella and independent Bella who had been freed for several generations. All were nomadic and at that time, although men had considerable contact with other populations through markets and trade, most women led socially restricted lives with little knowledge of the wider world.

Recent changes and crises

Drought in 1984 and 1985 led to substantial herd losses, population movements and a burgeoning of international and local NGO interventions. Many dependent Bella left their impoverished owners, people moved temporarily to the towns and some Tamasheq groups started to sedentarise (Randall and Giuffrida 2006). Those who remained nomadic became less isolated, with increased contact with wider Malian society and with development projects.

In 1990, a rebellion broke out in Niger and was followed by Tuareg attacks in east Mali on military and administrative posts. The MPLA (Mouvement Populaire pour la Libération de l'Azawad) was created with the aim of liberating Tuareg territories in the north and was countered by Malian army patrols. Rebel attacks increased in intensity throughout the early part of 1991 and gradually expanded westwards towards Timbuktu and the Mema with escalating retaliations by the Malian army on the Berber Tuareg and Maures with men, women and children being killed in camps, villages and towns. The Malian population became incited against the Tuareg and there were attacks and raids on shops owned by Tuareg and Maures throughout northern and central Mali. Physical appearance was a major factor identifying those who were attacked and in May 1991, Tuareg in the Delta and Mema areas fled en masse to Mauritania.[2] Some took their herds and tried to continue mobile pastoralism in Mauritania where they encountered major problems accessing water and wells. Others left everything behind or consumed most of their animals during the flight.

Three refugee camps were set up in south east Mauritania where conditions were poor at first because of the scale of the crisis and the isolation of the area. People continued to arrive into the refugee camps up until 1994 and the majority stayed until 1996, having spent four or five years there. Although spontaneous repatriations occurred throughout the period, the main movement out was in 1996 under a repatriation programme run by UNHCR (United Nations High Commissioner for Refugees) and GTZ (Deutsche Gesellschaft für Technische Zusammenarbeit or German Technical Cooperation), after the signing of various peace agreements. Although most camp residents had previously been nomadic pastoralists, there were also people who had sedentarised after the 1984 to 1985 drought, along with civil servants, teachers, traders, craftsmen and students. A few domestic Bella fled with their masters. Black Tamasheq were not persecuted and most Bella remained in Mali, some with the animals, some leaving the pastoral sector altogether. Nomadic pastoralist Tuareg experienced substantial lifestyle changes in the refugee camps, including being fixed in one place with large numbers of people from different social groups alongside the educated and those who had left pastoral sector and zone. Young people enjoyed a varied and active social life. Rudimentary healthcare provision developed into immunisation programmes, free health and maternity care. Latterly schools were set up in the refugee camps.

Repatriation and transformations

Repatriation made further changes to lifestyle. Part of the reconciliation and repatriation package developed by the Malian government with UNHCR and other international organizations (République du Mali 1995) included promises to build schools, drill boreholes and develop infrastructure in the specific destinations that refugees were obliged to name and return to as well as in other northern communities. For repatriated refugees infrastructure was to be proportional to the population registered. This encouraged sedentarisation and has led to a proliferation of wells surrounded by small settlements (Randall and Giuffrida 2006). Thus, in 2001, four years after repatriation, the population was much more sedentary, less dependent on a pastoral economy, and unpaid domestic labour was rarely available. Formal education was more acceptable and available but still uncommon and there was an increased use of modern health services.

Methods

Two single-round demographic surveys were undertaken in Mali in 1981 (the Delta, N=6125) and 1982 (the Gourma, N=6520), followed up by an intensive qualitative three-month study in one Tamasheq camp (Randall 1984, 1996) at a time when the western Kel Tamasheq population spent the dry season using pastures in the inner Niger Delta, leaving in the wet season to move north and west into drier areas, including the Mema (Figure 3.1). Those surveyed were all nomadic pastoralists practising no agriculture and were socially heterogeneous, including all the different Tamasheq social classes, warriors, religious maraboutic groups, vassals, lower status groups, blacksmiths, Bella slaves and descendents of freed slaves.[3] Age-reporting was poor and in 1981 there was evidence of underreporting of neonatal deaths and of births by older women.

In 2001, a further demographic survey (N=8272) was undertaken which covered much of the same population studied in 1981. All

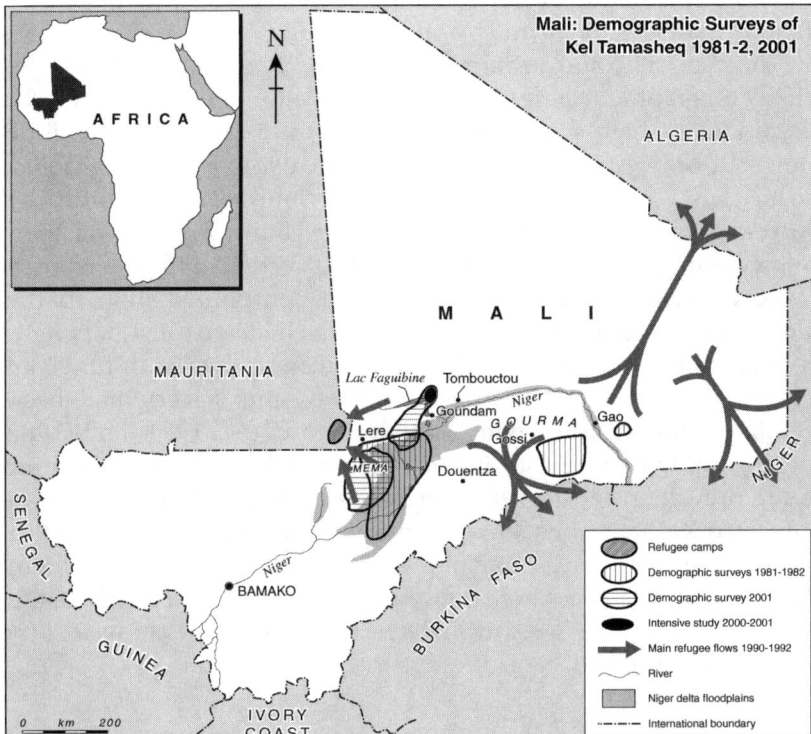

Figure 3.1: *Map of Mali indicating Tamasheq study sites in 1981–2001 and main refugee flows in the conflict of the 1990s*

the surveys used a household questionnaire and a birth history; 2001 also included a marital history for men and women and some questions on contraceptive knowledge. Alongside the 2001 survey, an anthropological study was undertaken in three northern communities which ranged in the extent of their sedentarisation and dependence on pastoralism.

Tamasheq demography in the 1980s

The 1980s surveys had shown the Kel Tamasheq to be demographically unusual for sub-Saharan African populations. Heterogeneity in terms of production, environment and social organization within the Malian Kel Tamasheq population means that we cannot generalise about their demography to other Kel Tamasheq populations in Mali: but some of the specificities almost certainly apply elsewhere.[4] The demographic regime was typified by low(ish) fertility,[5] largely a function of the nuptiality regime, and unusual patterns of mortality differentials (see below). Although extramarital childbearing was more acceptable for Bella, overall their total fertility was similar to that of Tuareg (Randall and Winter 1985).

The economic and social role of women had a major impact on the demographic regime (Randall 1984; Fulton and Randall 1988). Traditionally, high-status Tuareg women were expected to do little domestic or livestock-related work and the existence of Bella labour made this possible. Differences in behaviour were reinforced by force-feeding many high status girls and young women and their subsequent obesity limited their physical activity. Tuareg women were expensive to maintain and often contributed little to the household economy, housework and even childcare (see Chapter 4).

Nevertheless, there was substantial diversity over both time and space. The extent of both force-feeding and slavery had been declining for at least two decades before the 1981 and 1982 demographic surveys, but in the study populations both remained quite prevalent. In the 1980s, there was a small urban minority of educated Kel Tamasheq, but in both the populations studied in 1981 and 1982 everyone was nomadic, few (none of whom were women) had been to modern school and there was little contact with health services and camps were usually relatively small (twenty to fifty people) and isolated.

Tuareg obesity

In 1981, a substantial proportion of Tuareg girls and young women were extremely fat, having been force-fed from about age eight. Many older women were also obese, though often the sight of huge wrinkled sacks of skin under their arms was evidence of great weight in the past that had not been maintained. These women gave accounts of force-feeding in their youth which are very similar to those outlined by Walentowitz in Chapter 4, although these practices were said to have largely ceased during the 1973 drought. Accounts agreed that women were not as fat as they had been in the previous generation but, in 1981 in one large camp, there were several women whose weight approached 100 kg (Wagenaar-Brouwer 1985). Force-feeding was declining, largely because livestock loss as a result of drought, meant that many people no longer had the necessary resources. Women still wanted to be fat and men still desired very fat women. In 1982, I observed young women spending whole days with bowls of millet mixed with butter which they mixed into balls and then washed down with water, clearly forcing themselves to continue. Their stated aim was to get stretch marks on their thighs and stomach.

Tuareg obesity was obtained both through very substantial calorie intake: largely through forced consumption of huge amounts of milk, butter and grain, and through little energy expenditure. Tuareg women in the early 1980s were remarkably immobile, particularly considering the nomadic pastoral production system. Women spent a huge proportion of their time just sitting, either in their own tent, or with other women in a neighbouring tent or in the shade of a tree. They rarely moved far from the camp, and when they did walk anywhere it was extremely slowly and with a curious swinging gait (possibly because of their huge thighs and buttocks). When the camp moved, Bella women did all the packing; the Tuareg woman would just sit until her mount – a camel or an ox – was brought to her. Women's saddles were designed so that they could retain their habitual cross-legged pose in the saddle. On arrival at the new camp, the Tuareg women would sit in the shade of a tree whilst the Bella pitched the tent and cleared the ground. Most Tuareg women were observed to perform only two main types of work: they would shake the skins full of soured milk to make butter and they would make and repair the leather tents. Both of these activities were undertaken whilst in their habitual cross-legged pose.

Why were Tuareg girls force-fed?

There were many dimensions to Tuareg force-feeding, only some of which were articulated. Fat was beautiful and for men, fat women were highly sexually attractive. Young men would spend long days and evenings courting young girls, with the really obese being most sought after. Men frequently expressed a desire for fat women: as was also the case in the Arab population researched by Popenoe (2003) in Niger. But force-feeding both demonstrated wealth and was a means of storing excess production; wealth in cows who produced a surplus of milk used for the feeding,[6] and wealth in Bella. A family could only initiate force-feeding if there was a Bella woman to supervise it, and once a woman was married she could only maintain her weight if she had Bella women to do all the daily work: pitching and striking tent, cleaning around the tent, adjusting the tent in line with the changing angle of the sun during the day, fetching water, cooking and childcare. Thus, only households rich in livestock and female Bella labour could force-feed their girls: and the obese products were physical evidence of this wealth. Furthermore, some Tuareg girls were married when they were very young. Although data are not available to analyse age at marriage by weight or by force-feeding status, the impression was certainly that most girls who married very young had been force-fed. These marriages were often before puberty and the young girl would go and live with her in-laws although the marriage would not be consummated immediately. Consummation was supposed to wait until the girl was mature enough and had reached puberty, but I was told that girls who were force-fed reached puberty earlier than other girls (probably true given the hormonal evidence presented by Ellison (2001)) but also that the rolls of fat which developed could be interpreted as the development of breasts and consummation did not need to wait for menstruation. This does seem to have been only a minor dimension of force-feeding since by no means all fat girls married very young.

With the exception of noting that fat girls were likely to reach puberty earlier than thin girls, it was never suggested that force-feeding bore any relationship to fertility. Even comments about early puberty were expressed in terms of legitimating sexual relations and not of accelerating childbearing. In general, neither Tuareg women nor men in 1981 and 1982 were particularly pro-natalist. Children were desired and were seen as the natural consequence of sexual relations and marriage, but unlike many neighbouring agricultural populations infertile women were not

Carsten, J. 2000. *Cultures of Relatedness: New Approaches to the Study of Kinship.* Cambridge: Cambridge University Press.

Chodorow, N. 1978. *The Reproduction of Mothering.* Berkeley: University of California Press.

Cohen, L., D. Perales and C. Steadman. 2005. 'The O Word: Why the Focus on Obesity is Harmful to Community Health', *Californian Journal of Health Promotion*, 3(3): 154–61.

Collier, J. and M. Rosaldo. 1981. 'Politics and Gender in Simple Societies', in S. Ortner and H. Whitehead (eds), *Sexual Meanings: The Cultural Construction of Gender and Sexuality.* Cambridge: Cambridge University Press, pp. 275–329.

Coveney, J. 2008. 'The Government of Girth', *Health Sociology Review*, 17(2): 199–213.

Crossley, N. 2004. 'Fat is a Sociological issue: Obesity Rates in late Modern, "Body-conscious" Societies', *Social Theory and Health* 2: 222–53.

Dinnerstein, D. 1976. *The Mermaid and the Minotaur: Sexual Arrangements and Human Malaise.* New York: Harper and Rowe.

Douglas, M. 1984 (1966). *Purity and Danger: An Analysis of the Concept of Pollution and Taboo.* London: Routledge.

Escobar, A. 1999. 'After Nature: Steps to an Anti Antiessentialist Political Ecology', *Current Anthropology* 40(1): 1–30.

Evans, B. and K. Le Besco. 2001. *Bodies Out of Bounds: Fatness and Transgression.* Berkeley and Los Angeles: University of California Press.

Fabian, J. 1983/2002. *Time and the Other: How Anthropology Makes its Object.* New York: Columbia University Press.

Fausto-Sterling, A. 2005 'The Bare Bones of Sex: Part One – Sex and Gender', *Signs: Journal of Women in Culture and Society* 30 (2): 1491–527.

Finkler, K. 2000. *Experiencing the New Genetics: Family and Kinship on the Medical Frontier.* Philadelphia: University of Pennsylvania Press.

Franklin, S. 1997. *Embodied Progress: A Cultural Account of Assisted Conception.* London: Routledge.

――― 2003. 'Re-thinking Nature-Culture: Anthropology and the New Genetics', *Anthropological Theory* 3(1): 65–85.

―――, C. Lury and J. Stacey. 2000. *Global Nature, Global Culture.* London: Sage.

Gard, M. and J. Wright. 2001. 'Managing Uncertainty: Obesity Discourses and Physical Education in a Risk Society', *Studies in Philosophy and Education* 20(6): 535–49.

――― and J. Wright. 2005. *The Obesity Epidemic: Science, Morality and Ideology.* London: Routledge.

Goffman, E. 1968. *Stigma: Notes on the Management of Spoiled Identity.* Middlesex: Penguin.

Gremillion, H. 2005. 'The Cultural Politics of Body Size', *Annual Review of Anthropology* 34: 13–32.

Haraway, D. 1989 *Primate Visions: Gender, Race and Nature in the World of Modern Science*. London: Routledge.

Heslehurst, N., R. Lang, J. Rankin, J.R. Wilkinson and C.D. Summerbell. 2007. 'Obesity in Pregnancy: A Study of the Impact of Maternal Obesity on NHS Maternity Services', *BJOG: An International Journal of Obstetrics and Gynaecology* 114(3), 334–42.

Ingold, T. 1988. 'The Animal in the Study of Humanity', in T. Ingold (ed.), *What is an Animal?* London: Unwin Hyman.

Kahn, S. 2000. *Reproducing Jews: A Cultural Account of Assisted Conception in Israel*. Durham, NC: Duke University Press.

Kwan, S. 2006. 'Contested Meanings about Body, Health, and Weight: Government, Activists, and Industry Framing Competitions over the Overweight Body'. Paper presented at the American Sociological Association Meetings, August 2006.

Labinger, J. and H. Collins (eds). 2001. *The One Culture? A Conversation about Science*. Chicago: University of Chicago Press.

Latour, B. 1993. *We Have Never Been Modern*, trans. C. Porter. London: Harvester Wheatsheaf.

Lewin, E. 1995. 'On the Outside Looking In: The Politics of Lesbian Motherhood', in F. Ginsburg and R. Rapp (eds), *Conceiving the New World Order: The Global Politics of Reproduction*. Berkeley: University of California Press, pp. 103–21.

Longhurst, R. 1995. 'Fat Bodies: Developing Geographical Research Agendas', *Progress in Human Geography* 29(3): 247–59.

Lucas, R. and R.J. Barrett. 1995. 'Interpreting Culture and Psychopathology: Primitivist Themes in Cross-Cultural Debate', *Culture, Medicine and Psychiatry* 19(3): 287–326.

MacCormack, C. and M. Strathern (eds). 1980. *Nature, Culture and Gender*. Cambridge: Cambridge University Press.

McBride, A. 1989. 'Fat is Generous, Nurturing, Warm … ' *Women and Therapy* 8(3): 93–103.

Monaghan, L. 2005a 'Discussion Piece: A Critical Take on the Obesity Debate', *Social Theory and Health* 3: 302–14.

———— 2005b 'Taking Sides: Can Being Fat be Good For You? Yes', *Men's Health Forum* 7: 8-9.

———— 2007. 'Body Mass Index, Masculinities and Moral Worth: Men's Critical Understandings of "Appropriate Weight-for-Height" ', *Sociology of Health and Illness* 29(4): 584–609.

Mullin, M. 1999. '*Mirrors and Windows: Sociocultural Studies of Human-Animal Relationships*', *Annual Review of Anthropology* 28: 201–24.

Murray, S. 2008. *The 'Fat' Female Body*. London: Palgrave, MacMillan.

Neumark-Sztainer, D. 1999. 'The Weight Dilemma: A Range of Philosophical Perspectives', *International Journal of Obesity* 23 (Suppl 2): 31–37.

Orbach, S. 1978. *Fat is a Feminist Issue*. London: Arrow Books.

reviled or necessarily divorced and women did not acquire status through being the mother of many children. A woman was born into a social class and her status was obtained through her birth position, her kin, her behaviour and, to an extent, her obesity. Thus, there is no evidence that fattening was intended to enhance reproduction, nor that people considered that it inhibited fertility.

Health consequences of obesity

The consequences of obesity were quite far-reaching although it is difficult to disentangle some of the physical sequelae from the behavioural patterns generated by cultural values which contributed to the obesity in the first place. No health data are available for this population and in 1981 and 1982 few accessed modern health services. They did have a good understanding of anatomy and physiology from their close relationships with livestock and there was an extensive and well developed traditional medicine. Indirect estimates of child and adult mortality show an extraordinary pattern of mortality whereby high status, rich Tuareg women had a lower adult life expectancy than low status Bella women; Tuareg children had higher infant and child mortality than Bella children but Bella men had lower adult life expectancy than Tuareg men (Randall 1984; Hill and Randall 1984; Hilderbrand et al. 1985). Generally, demographers anticipate that children of richer, higher status social subgroups will have lower mortality than the poorer subgroups unless there is a major difference in breastfeeding behaviour: which here there was not. Explanations for these child mortality differences are rooted in both the obesity of many Tuareg women and also in the normative, immobile behaviour expected from Tuareg women whether obese or not.

A life of physical immobility and the delegation of most work (cooking, domestic and childcare) to Bella girls and women was normal for those women who had access to Bella labour whether they had been force-fed or not.[7] In a hazardous environment (highly contaminated standing water and high prevalence of many infectious diseases, especially malaria and diarrhoeal diseases) with no access to health care or immunisations, a child's survival was largely predicated upon the daily care and attention it received in terms of washing, availability of food and water and general caring input. Observations suggested that young Bella children who spent most of the day with their mothers, were washed more regularly and received more continuous care, and were generally looked after better than Tuareg children who might be cared for partially by their mothers (who did breastfeed their children) and partly by young (often aged six or seven) Bella nursemaids who were unable to provide adequate and continuous care. One obese Tuareg woman admitted that she felt she should pay more attention to her young child but that the heat and the effort involved in moving just made it easier to delegate to Bella nursemaids.

The fact that Tuareg women had lower adult life expectancy than Bella women indicates some of the health costs of obesity. Women were always complaining about their health and particularly about their heart and various internal problems. Many medical studies have shown that obesity increases the risk of cardio-vascular disease, diabetes and other chronic diseases and Tuareg women also experienced these chronic disease problems. On top of that, there are also risks to both mother and child that are associated with obesity during pregnancy and childbirth. Recent studies in Sweden, Wales and Finland (Cedergren 2004; Usha Kiran et al. 2005; Raatikainen et al. 2006) show that obese women are much more likely to suffer from a range of detrimental pregnancy outcomes including preeclampsia, fetal death, induced delivery, caesarean section and perinatal death. Given that Tuareg women had no access to health services, increased risk of preeclampsia and the preconditions which would indicate induction or caesarean delivery will also have increased maternal and perinatal mortality on top of the baseline increased risk of fetal and perinatal mortality. From data on survival of sisters in the 2001 survey the estimated lifetime risk of maternal mortality was one in eight: much higher than that of most of the Sahelian national levels (WHO 2005, Annex 3). These estimates applied to the preceding twenty-five years including the period when many more women were obese.

Breastfeeding consequences of obesity

The literature on obesity and breastfeeding concentrates primarily on issues around whether breastfeeding inhibits the subsequent development of obesity in the breastfed child. Few studies investigate whether obesity has an impact on women's ability to breastfeed: and those that do tend to focus mainly on issues of socioeconomic correlates and body image which might reduce both initiation of breastfeeding and duration in Western populations. Neither of these are relevant in the Tuareg case where all women breastfeed for nearly two years unless a pregnancy intervenes or they are completely unable to do so.[8] Body image is not an issue with respect to breastfeeding in this population but there remains the issue of

Pashigian, M. 2002. 'Conceiving the Happy Family: Infertility and Marital Politics in Northern Vietnam', in M. Inhorn and F. van Balen (eds), *Infertility Around the Globe: New Thinking on Childlessness, Gender and Reproductive Technologies.* Berkeley: University of California Press, pp. 134–51.Rabinow, P. 1996. *Making PCR: A Story of Biotechnology.* Chicago: Chicago University Press.

Rheinberger, H. 2000. 'Beyond Nature and Culture: Modes of Reasoning in the Age of Molecular Biology and Medicine', in M. Lock, A. Young and A. Cambrosio (eds), *Living and Working with the New Medical Technologies: Intersections of Inquiry.* Cambridge: Cambridge University Press, pp. 19–30.

Rich, E. and J. Evans. 2005. ' "Fat Ethics" – the Obesity Discourse and Body Politics', *Social Theory and Health* 3(4): 341–58.

Roepstorff, A. 2001. 'Thinking with Animals', *Sign Systems Studies* 29(1): 203–18.

———, N. Bubandt and K. Kull. 2003. *Imagining Nature: Practices of Cosmology and Identity.* Aarhus: University Press.

Ruddick, S. 1980. 'Maternal Thinking', *Feminist Studies* 6: 342–64.

Saguy, A. and K. Riley. 2005. 'Weighing Both Sides: Morality, Mortality, and Framing Contests over Obesity', *Journal of Health Politics, Policy and Law* 30(5): 869–923.

——— and R. Almeling (2008). 'Fat in the Fire? Science, the News Media and the "Obesity Epidemic" ', *Sociological Forum* 23(1): 53–83.

Said, E. 1979. *Orientalism.* New York: Vintage Books.

Sobal, J. 2004. 'Sociological Analysis of the Stigmatisation of Obesity', in J. Germov, J. and L. Williams (eds), *A Sociology of Food and Nutrition: Introducing the Social Appetite* (second edn.). Melbourne: Oxford University Press, pp. 383–402.

Stewart, P., and A. Strathern. 2001. *Humors and Substances: Ideas of the Body in New Guinea.* Westport CT/London: Bergin and Garvey.

Strathern, M. 1988. *The Gender of the Gift: Problems with Women and Problems with Society in Melanesia.* Berkeley and Los Angeles: University of California Press.

——— 1992. *After Nature: English Kinship in the Late Twentieth Century.* Cambridge: Cambridge University Press.

——— 1999. *Property, Substance and Effect: Anthropological Essays on Persons and Things.* London: Athlone.

Teman, E. 2003. 'The Medicalisation of "Nature" in the "Artificial Body": Surrogate Motherhood in Israel', *Medical Anthropology Quarterly* 17(1): 78–98.

Thomson, R and T. Garland. 1996. *Freakery: Cultural Spectacles of the Extraordinary Body.* New York/London: New York University Press.

Torgovnick, M. 1990. *Gone Primitive: Savage Intellects, Modern Lives.* Chicago: University of Chicago Press.

Upton, R. and S. Han. 2003. 'Maternity and its Discontents: "Getting the Body Back after Pregnancy', *Journal of Contemporary Ethnography* 32(6): 670–92.

Van Dijck, J. 2002. 'Medical Documentary: Conjoined Twins as a Medical Spectacle', *Media, Culture and Society* 24, (4): 537–56.

Warin, M., K. Turner, V. Moore and M. Davies. 2008. 'Bodies, Mothers and Identities: Rethinking Obesity and the BMI', *Sociology of Health and Illness* 30(1): 97–111.

Zivkovic, T., M. Warin, M. Davies and V. Moore. 2010. 'In the Name of the Child: The Gendered Politics of Childhood Obesity', *Journal of Sociology* 40(4): 375–92.

Chapter 3

FAT AND FERTILITY, MOBILITY AND SLAVES:
LONG-TERM PERSPECTIVES ON TUAREG OBESITY AND REPRODUCTION

Sara Randall

Introduction

The perceived relationships between fatness and fertility differ substantially over time and space. Many prehistoric depictions of fertility show plump female figurines and it is clear that fatness was seen as evidence of reproductive potential. Evolutionary theory predicts that thinner women will reproduce more slowly because they do not have adequate resources to invest in pregnancy and a range of studies have demonstrated that maternal nutritional and energetic status plays some role in the probabilities of conception, taking a pregnancy to term and producing a child who is likely to survive (Ellison 2001). On the other hand, research has shown that human females are remarkably resilient to malnutrition in pregnancy and are able to take a child to term under nutritional stresses that most other mammals are unable to tolerate (Durnin 1987; Prentice and Prentice 1988) and that human lactation remains possible even when mothers are energetically stressed (Prentice et al. 1994). However, in populations where fatness and obesity are socially undesirable, fat women may find it harder to find partners and thus reproduce and the impact of obesity on fertility appears to be largely negative via pathways of social selection. There is also evidence that obesity may be linked with increased infertility (Gesink Law et al. 2007) and that obese women may have poorer pregnancy outcomes and higher fetal and neonatal mortality (Cedergren 2004; Raatikainen et al. 2006).

Much of the research on nutritional stress and reproduction has been undertaken in the Sahel (mainly in the Demographic Surveillance Systems in the Gambia and Senegal) where strong seasonality, poverty and a dependence on rain fed agriculture and women's labour provide a context where we can learn much about human adaptation to malnutrition and food shortages. Given this emphasis and a general association of nomadic lifestyles with mobility and activity, it might seem surprising that a Sahelian nomadic pastoralist population could also be associated with obesity. Yet there are several such populations in a band which stretches from Mauritania across Northern Mali into Niger (*New York Times* 2007; Popenoe 2003; Chapter 4) and includes Maures and some Tuareg and Arab populations.

Here, I examine diverse ideas and evidence relating fatness, fertility and reproduction over recent decades for a Malian Tuareg population, ranging from an examination of the representations of the relationship by colonial administrators through to data from demographic surveys in the early 1980s and in 2001. Using these sources, I consider whether there is, indeed, any evidence for a direct or indirect association between female fatness and reproduction for this population and the different pathways through which this relationship might operate.

Malian Tuareg

The Kel Tamasheq (or Tuareg) live across Northern Mali, southern Algeria, Niger and northern Burkina Faso. Many used to be archetypal nomadic pastoralists, herding goats, sheep, cattle and camels according to the local environment.

Tuareg social class

In these Tamasheq populations, the higher status social classes are descended from Berber populations who crossed the Sahara, in the fifteenth and sixteenth centuries. Tamasheq is a Berber language and physically most higher-status Kel Tamasheq (known as Tuareg) are of Berber origin.[1] As in many West African communities, slavery of context-based literature on obese bodies, and an ethnographic approach enabled us to focus on the culturally situated knowledge and practices of participant's everyday lives. We conducted eighteen months of fieldwork (from 2004 to 2005) with the women in their own environments, where we met them in their own homes and neighbourhoods. Every day we accompanied them on grocery outings, mapped local precincts (including location of shops, schools, public transport, exercise facilities and green spaces) and joined them on exercise programmes. Each participant was interviewed at least twice (some three times) during which we explored their memories of food and weight growing up, the embodied experiences of eating and being large (including pregnancies), meanings and symbolic language of food, and the relationship of these experiences to motherhood. In addition, we collected genealogies as we were interested in what discourses (if any) these women drew upon such as hereditary claims to weight, as well as wider symbolic connections of kinship and food. Without prompting, participants often spoke of their mothers and grandmothers bodies, comparing their own bodies to generations of bodies that changed across life-courses. As well as traditional kinship ties, these women also spoke about the different ways in which food was used symbolically (often by grandparents) to 'spoil' and nurture them. These genealogies and their contextual information provided a rich source of explanation concerning reproduction of body shapes over successive generations and the social relationships around which food, gender and bodies pivot.

Interviews were primarily conducted by the first author of this chapter and were usually two to three hours each in length for each sitting. They were conducted mainly in kitchens and lounge rooms, and, along with field notes, were transcribed verbatim, and indexed with a qualitative assisted computer package (NVivo). As with most qualitative work, we followed an inductive approach to analysis in which we searched for patterns, themes and disjunctures, in order to position the data in light of the empirical evidence and current theoretical approaches to embodiment.

The thirty women were a heterogeneous group, and ranged in age from twenty-three to forty-four years, and came from differing socioeconomic and ethnic backgrounds. It is important to note that there were class, age and ethnic differences amongst these women, and that definitions of motherhood, pregnancy, gender, weight and food varied accordingly. In a previous paper (Warin et al. 2008), we have discussed gendered and class-based experiences of embodiment and obesity; however, this chapter does not use these categories as analytic departures. Our analysis here focuses on implicit assumptions about 'nature', and how and when it is called upon to explain these women's understandings of body size.

We did not enter this research assuming that women would agree with or self-identify as overweight or obese. Also, we were aware of the sensitivity of a topic that Carryer (1997: 99) describes as 'private and often extremely painful' experience. For these reasons, we did not have the word obesity in our project title, but were clear that our project was concerned with women, food and bodies.[2] In all of our interactions we listened carefully to how participants described their own bodies, taking their language of representation as our cue. It was only then could we ask about obesity, and their conceptualisations of what this meant. Other studies have also noted the difficulty of working with people labelled as obese (Longhurst 1995; Heslehurst et al. 2007), as the 'O' word is so heavily laden with individual culpability (Cohen et al. 2005).

Two of the women were pregnant at the time of interview and others had children ranging in age from one to seventeen. All but three of the participants described being 'big as a child', either as a young child or during adolescence (due to 'puberty' and 'hormonal changes'). Being bullied at school was a common feature of narratives due to body size. Other 'phases in life' where weight had fluctuated or was gained included pregnancies, 'getting married' and what one woman referred to as 'stress-related eating' (such as a child involved in a car accident, a family member diagnosed with serious illness, moving countries, and giving up smoking). There were often decisive moments when the women knew they had become too large, through disparaging looks or comments from strangers, for example. For Abby, it was going to her nephew's twenty-first birthday party and comparing herself to the entertainment; 'when the fat-o-gram took her clothes off I thought "God, she doesn't look much bigger than me" … that scared me … that was a wake up call. I could be a model for a fat-o-gram'

All of the women had engaged in various practices in attempts to lose weight. They had been on diets (including *Weight Watchers, Jenny Craig, Gloria Marshall, Herbalife* and 'fad' diets), one had had bariatric surgery (and another was on the waiting list) and one spoke at length about a longstanding eating disorder in her early twenties. Exercise had trailed off after leaving school, and was difficult to sustain as a parent (and particularly for single and time-poor mothers). Indeed, exercise (and especially for those on low incomes) was jokingly characterised as 'running after the kids, cleaning and scrubbing floors'.

as biology, as a shared genetic link and women's natural capacity for childbirth and nurturing (Carsten 2000: 7). As such, this argument presents a new theoretical approach to obesity that may provide opportunities for interdisciplinary thinking and action beyond current rhetorical imperatives.

Following a description of the study and the mothers involved, we outline how nature is used to situate obesity as 'not me'. By aligning obese bodies as closer to nature through the trope of Otherness, participants distanced obesity and rejected the clinical associations. When talking about their own self identifications, nature relied on a taken-for-granted, positive framing of the 'naturalness' of maternal bodies, in which reproduction, mothering and large bodies were positively valued. In concluding, we reflect on the importance of understanding biology-in/as-culture for those engaged in the 'obesity wars', suggesting that working within (rather than outside) competing epistemological frames of nature/culture could provide new analytical avenues of understanding obesity.[1]

The study

The thirty women involved in this study were sampled from a larger, ongoing epidemiological study located in Adelaide, Australia, which is investigating women's health during pregnancy and the growth and development of subsequent children. In the larger cohort of five hundred and fifty women, the high prevalence of overweight and obese women was noted: an observation which prompted this investigation. Similarly, a recent study in the north-east of England was also initiated by anecdotal observations of maternity and health care professionals who noted a rise in the number of obese expectant mothers in their health clinics (Heslehurst et al. 2007).

In the main cohort the participating women had their Body Mass Index (BMI) recorded. The first thirty women who fulfilled the clinical criteria of obese were purposively selected from the existing data base and invited (by letter) to participate. Six women declined to participate (stated reasons for refusal included working full time, busy with new baby, house renovations and illness). Women who declined were replaced by the next eligible person.

In order to understand how these mothers experienced their bodies and changing shapes, and the meanings and practices they gave to their everyday worlds, we employed ethnographic techniques of participant observation and in-depth interviews. There is a paucity constructed and socially produced (Escobar 1999: 2). Yet recent developments in genetic sciences and the associated work on kinship, relatedness and new reproductive technologies have 'convincingly argued that the social facts versus biological facts distinction is increasingly inadequate to describe the context of new genetics, as the biological is literally being rebuilt in ways that make it more social than "natural"' (Franklin 2003: 65). From this ethnographic work there are now different ways in theorising biological and social connections. Some have called for a more subtle analysis into the ways in which concepts of nature and culture travel, connect, disconnect and contain (Strathern 1992), invert (Rabinow 1996) or intensify each other (Rheinberger 2000). It is in this recent approach to the analytic puzzle of nature (and particularly Franklin's work into new genetics (2003)) that this argument is situated. Franklin's rethinking of nature and culture feeds directly into the key issues of this chapter; of the politics of mothering and reproduction, embodiment, kinship and relatedness. Like Franklin, we argue that the nature/culture opposition needs reinventing, as what is 'natural' is, in Latour's terms, both real and constructed. Instead of advocating for the abolition of the nature/culture dichotomy, we highlight how nature is used in everyday articulations and practices to make sense of large bodies and social relations. The work of Franklin et al. (2000, 2003) has surprising analogies with our endeavour and succinctly sums up our aim:

> By analysing the *traffic in nature*, we suggest that 'while nature and culture are increasingly isomorphic, in that they are acquiring each others' powers, their distinctiveness continues also to remain crucial.' We argue that the category of the natural remains central to the production of difference, not only as a shifting classificatory category, but through processes of naturalisation, denaturalisation, and renaturalisation … Following Strathern, we ask how the 'uses of nature as models for context' are shifting (Franklin et al. 2000: 20), and what effects, entities and embodiments are generated as a result. (Franklin 2003: 68)

The contexts of new reproductive technologies (and most particularly the remaking of nature through the artifice of technology) and obesity are of course different, but we borrow Franklin's metaphor (taken from Haraway 1989) of 'trafficking nature' to analyse how obesity is distanced and large bodies rationalised. This trafficking includes different idioms of nature; as in what animals do (and their relationship to humans), and nature whether obesity has a biological impact on lactation. An Australian study (Donath and Amir 2000) showed that obese women were less likely to initiate breastfeeding and breastfed for shorter duration even after control for maternal smoking, age and socio-demographic characteristics. They review the limited literature on experimental animals which suggests that there may be some biological relationships but conclude that it is largely the sociocultural and mental health characteristics of obese Australian women that determined breastfeeding behaviour. Obese Tuareg women did not suffer low self esteem or poor mental health given the positive cultural value of fatness. Nevertheless many Tuareg women who were (apparently) successfully breastfeeding complained that they did not have enough milk. The reality of this perception is hard to judge since, for a woman who has been force-fed with abundant milk and for whom large quantities of milk are a sign of largesse and wealth, it is quite possible that their perceptions of 'adequate' milk are misleading.

Fertility consequences of obesity

In a premodern Sahelian environment where most food was grown, gathered or raised, and where, in most populations, girls and women worked hard, it is likely that there was considerable childhood malnutrition and general seasonal nutritional stress (Chambers et al. 1981) and that age at menarche was late. In such a context the force-fed Tuareg girls who consumed a high fat and protein diet probably reached menarche earlier and could initiate childbearing earlier than girls from neighbouring populations (Ellison 2001).[9] Evolutionary ecological approaches to reproduction predict that under-nutrition and weight loss impact negatively on fertility, not over-nutrition. However, recent medical research shows that women with fertility problems are more likely to be either obese or very underweight and that fecundity is somewhat reduced for overweight and obese women (Gesink Law et al. 2007) with overweight women taking a median one month longer to conceive and obese women two months longer than those with an optimal BMI (Body Mass Index). In a non-contraception using population such as the Tuareg such delays would lead to slightly lower fertility but the impact would be minimal.

Social consequences of obesity

The ramifications of the value of fatness for women in Tuareg society are substantial, with implications for women in general; their roles, social lives, marriages, and thereby fertility and reproduction. From the outsider's perspective these overweight Tuareg girls and women led lives that were extremely circumscribed and inhibited by their immobility. Social prestige attached to this immobility stretched beyond those women who were very fat, to other women who were not, but usually aspired to being, fat. The high value of immobility and inactivity rendered women expensive commodities. Unlike most other traditional African communities where women were valued for their labour and their contribution to the household economy as well as for their reproduction, Tuareg women were a major cost: they cost money to marry, through the bridewealth, and although some brought cattle and Bella with them into marriage, they also cost a lot to maintain. They expected gifts whilst being courted as unmarried girls and gifts as married women, not to mention the costs of feeding them. It was believed that if a pregnant woman were denied something that she really wanted, she would suffer a miscarriage. This presented further costs for husbands, responding to her requests to reduce the chance of miscarriage.

Obesity facilitated the control of women by men. Any movement or visits were dependent on men because they needed transport managed by men. Very fat girls were objects of sexual attraction waiting to be courted by men (but not necessarily married). As symbols of familial wealth and status, their obesity demonstrated that they were submissive and conformed to social expectations and fatness also restricted their movement and independence making them highly desirable wives. Women had few opportunities for any personal development save through consolidating kinship links through marriage and childbearing. When doing research with Tuareg, men often expressed disbelief when we wanted to talk with women, claiming that most women knew nothing and were totally ignorant. Although this was not entirely true there was certainly an extent to which their immobility hindered their acquisition of knowledge about the wider world. They had to wait until that knowledge came to them which, if it did, was through male visitors. Most women had extremely limited personal aspirations and their lives largely revolved around discussions of marriage and kinship.

Marriage was the domain where fatness and fertility became closely interrelated. Most marriages were arranged between close kin; in the years preceding 2001 about half of first marriages were between first cousins whether cross or parallel cousins (Randall 2005). Since all Tuareg reproduction occurs within marriage, the monogamous marriage pattern is the major constraint on population level fertility.

Monogamy combined with substantial spousal age differences means that substantial proportions of reproductive aged women are unmarried at any one time, and marital mobility is facilitated by frequent and stigma-free divorce. Anything that influences Tuareg marriage influences Tuareg population fertility rates.

Tuareg fertility and fatness

Different interest groups at different times present contrasting perceptions of Tuareg fertility, fertility dynamics and the role of obesity in this fertility. Three perspectives will be discussed here: the attitudes of the French colonial administrators, evidence from the demographic surveys undertaken in 1981, 1982 and 2001, and the perceptions of the Tuareg population themselves as understood through observations in 1981 and 1982, and anthropological fieldwork in 2000 and 2001.

French colonial administrators

French administrators' attitudes to Tuareg fertility can be gleaned from an examination of the archives. Many political reports in the 1920s and 1930s comment on the slow or even negative Tuareg population growth as assessed through the censuses undertaken for tax purposes. It is very possible that the French colonial interpretations of Tuareg demographic dynamics were faulty (Randall 2009) but nevertheless they showed strong opinions about both the practice of force-feeding and its relationship with fertility. They were also highly critical of Tuareg marriage patterns which they felt inhibited fertility, and they disapproved of the high bridewealth. The following are examples of French colonial representations of Tuareg demographic dynamics and the role of fattening (my translations from IE 16 (Gourma Rharous) and IE 18 (Goundam)):

> The Tuareg themselves look like they are disappearing and in some time in the future only the Bellah (captive) element of the population will remain. The Tuareg have few children, in contrast the captives are prolific. It seems that the systematic fattening of Tuareg women and the stoutness they achieve is one of the main causes of their low birth rates. (IE 16: 1930)

> The decline of the white Tuareg population has been observed for a long time. For the Imushar, apart from the excessive bride price, this decline is explained by the fattening of the women which often makes

them sterile and by the marriage of pre-pubertal girls. I have seen children of 10 or 12 married and already living in the (adult) husband's tent. This practice accentuates the problems posed by fattening and increases the proportion of sterile women. (IE 18: 1933)

Research on the causes of the decline of the white population ... the principal barrier to births is the excessive bridewealth claimed by the women ... spoilt households of old men married to young girls of 10–15 are common. Normal households of young couples are very rare and it is easy to understand that a population whose reproduction depends on old men doesn't increase and even declines. (IE 18: 1933)

The Tuareg race seems to be clearly declining despite the efforts made to encourage marriage and forbid the fattening of women. (IE 18: 1942)

This low [population] growth over quite a long period must be due to two causes:
(1) High levels of stillbirth and numerous miscarriages due to the physical state of the women [i.e., their obesity].
(2) Late male marriage because of the impossibility of acquiring the required bridewealth before a certain age. (IE 18: 1943)

In fact, most comments in the archives suggest that the French administrators found many aspects of Tuareg behaviour with respect to sexuality, marriage and reproduction morally repugnant and counter to what should be happening in a 'normal' population. They even went to the extent of trying to ban force-feeding, control bridewealth and promote marriage through propaganda and provision of gifts.

French administrators particularly disliked marriages between older men and very young girls. Such marriages still occurred in 1981 and 2001 and were usually marriages arranged to develop particular kinship links; they often ended in the girl running away and divorce. It is likely that, although marriages between older men and young girls did exist, they were not as frequent as the French administrative reports suggest – but that they were the marriages that were noticed – primarily because they were the ones which were most unacceptable to the young male French administrators.

Demographic data

Prior to our demographic surveys undertaken in 1981and 1982 there were no data on Malian Tuareg demographic dynamics except for the French colonial administrative accounts and censuses.[10] So, it was hardly surprising that the general consensus in academic literature was that the Tuareg had low fertility (Swift 1977) and consequently low population growth rates. The overall picture in 1981 and 1982 of these two Tamasheq populations was a mean-reported parity at the end of childbearing of between five and six (see Figures 3.2 and Figure 3.3), which was lower than other Malian populations, including Bambara farmers surveyed as part of the same series of demographic surveys (see Table 3.1).

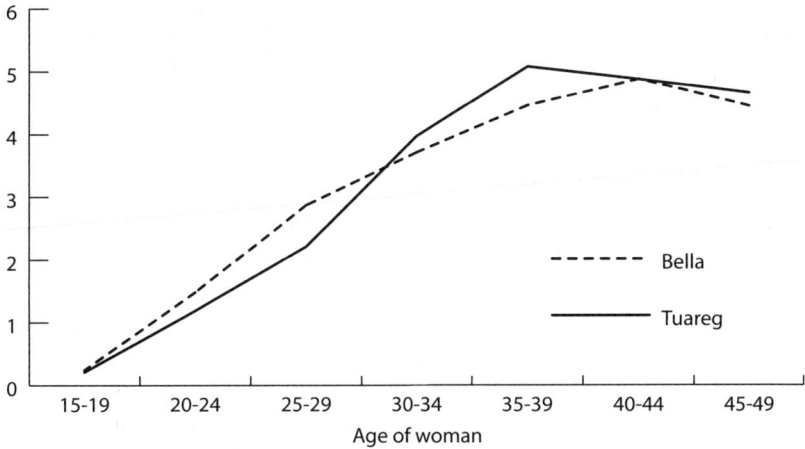

Figure 3.2: *Delta Tamasheq 1981 – reported mean parity by age*

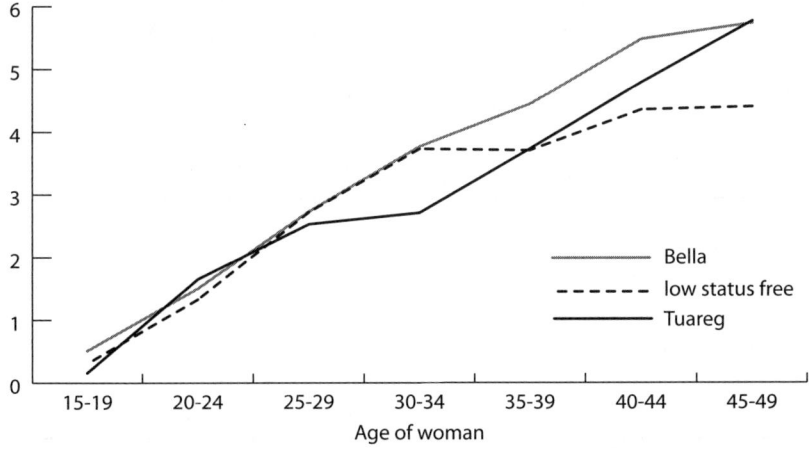

Figure 3.3: *Gourma Tamasheq 1982 – reported mean parity by age*

Table 3.1: *Tamasheq and Malian fertility in the early 1980s.*

	TFR	Completed Reported Parity (Women aged 40 to 49)
Mali 1981–83	7.10	7.06
Rural Mali 1981–83	7.17	7.01
Mopti, Gao, Timbuktu 1981 to 1983 (excluding nomads)	7.32	7.63
Bambara farmers 1981	8.1	7.31
Delta Tamasheq 1981	5.9	4.72
Gourma Tamasheq 1982	5.1	5.02

Source: Mali data: République du Mali 1987; Tamasheq and Bambara, author's own data

Total fertility rates estimated using the P/F method showed that Tuareg and Bella had similar overall fertility (Table 3.2) even if they achieved it in different ways (see Randall and Winter 1985). Overall, however, the relatively low Tamasheq fertility was achieved primarily through the marriage regime. No one used contraception, breastfeeding was similar in terms of duration and intensity throughout rural Mali, but the marriage patterns were very different: with the monogamous Tuareg women spending substantial proportions of their reproductive years outside marriage because there were not enough available men of marriageable age. In the Gourma, there is some evidence that fertility was also being depressed by something else (see TMFRs in Table 3.2) which could have been sexually transmitted diseases (Randall 1996) but is unlikely to have been obesity: Tuareg women in the Delta were as fat, if not fatter, than those in the Gourma and obesity was more prevalent, and also the 'low status free Tamasheq' in the Gourma

Table 3.2: *Tamasheq total fertility rate* and total marital fertility rate.*

	TFR	TMFR
Delta Tuareg 1981	5.9	10.0
Delta Bella 1981	5.9	9.6
Gourma Tuareg 1982	5.2	7.7
Gourma Low status free 1982	5.1	7.2
Gourma Bella 1982	5.1	8.7

* Estimated using P/F.
Source: Author's own data

have the lowest marital fertility and they were neither force fed nor
did they have access to Bella labour.

The 2001 demographic survey covered largely the same
population as the 1981 Delta survey. In the interim much had
changed: substantial herd loss, conflict, refugee camps, repatriation
and sedentarisation. Younger women were no longer force-fed[11]
and many were thin. The marriage pattern had barely changed
(Figure 3.4) with similar proportions of reproductive-age women
unmarried at any one time (Randall 2005).

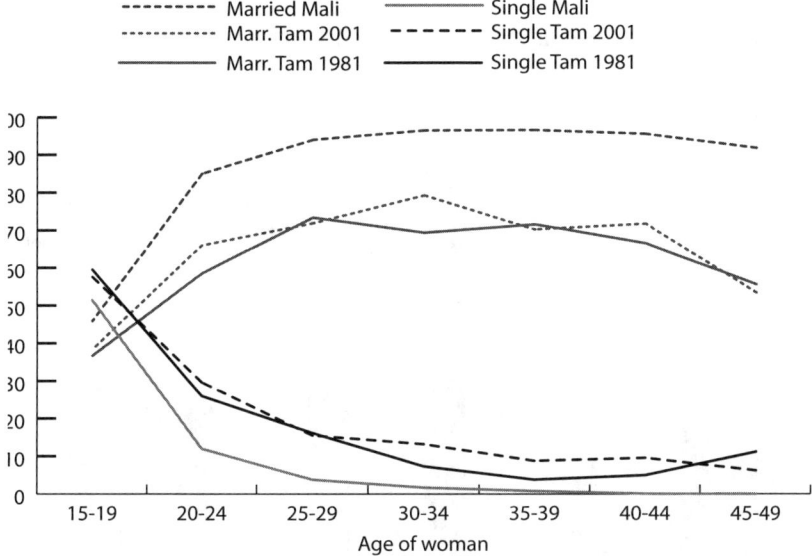

Figure 3.4: *Percentage Tamasheq women currently single and married,
1981, 2001 and Mali 2001 (DHS)*

Better data on age and age at first marriage in 2001 showed that
about one-fifth of women first married before they were fifteen,
although in the vast majority of those marriages the girls were
fourteen (see Figure 3.5). During the conflict the pattern of age at
first marriage had changed slightly but after repatriation it reverted
back. Thus marriage of very young girls observed by the French
administrators persisted in the 1980s and 1990s. Some of these girls
were married to much older men: but such marriages were not the
most frequent form (see Figure 3.6) and by the late 1990s they could
not have been related to force-feeding because the practice had
stopped. These marriages were, however, still arranged by kin and

were seen to be advantageous to the wider kin group. Other changes were occurring. In the 1970s, girls would be unlikely to refuse to marry, although they might run away after the marriage and many marriages broke down very quickly. By 2001, after repatriation, girls and young women were developing more control over their own marriage and it was not uncommon for them to refuse one or more marriages suggested by their kin. This new autonomy was not necessarily a direct function of the cessation of force-feeding but it was certainly part of the new freedoms of physical mobility, autonomy and independence which were fostered by the changing conditions in the refugee camps and part of which was an abandonment of the ideal of force-feeding.

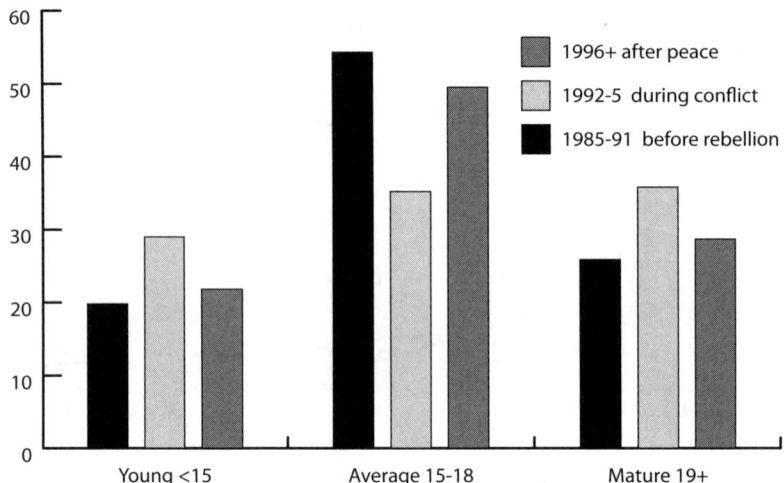

Figure 3.5: *Percentage distribution of age at first marriage of Tamasheq women by time period*

From the demographic data there is no evidence of a relationship between Tuareg obesity and their fertility. Over the twenty year period between 1981 to 2001 obesity from force-feeding had largely disappeared. A comparison of reported parity by age[12] for 1981 and 2001 (see Figure 3.7) shows effectively no change at all in fertility for the five younger age groups.[13] Not only did the fertility regime remain unchanged but marriage remained the most important proximate determinant of fertility, surprisingly little altered by all the socio-economic transformations occurring within Tuareg society.

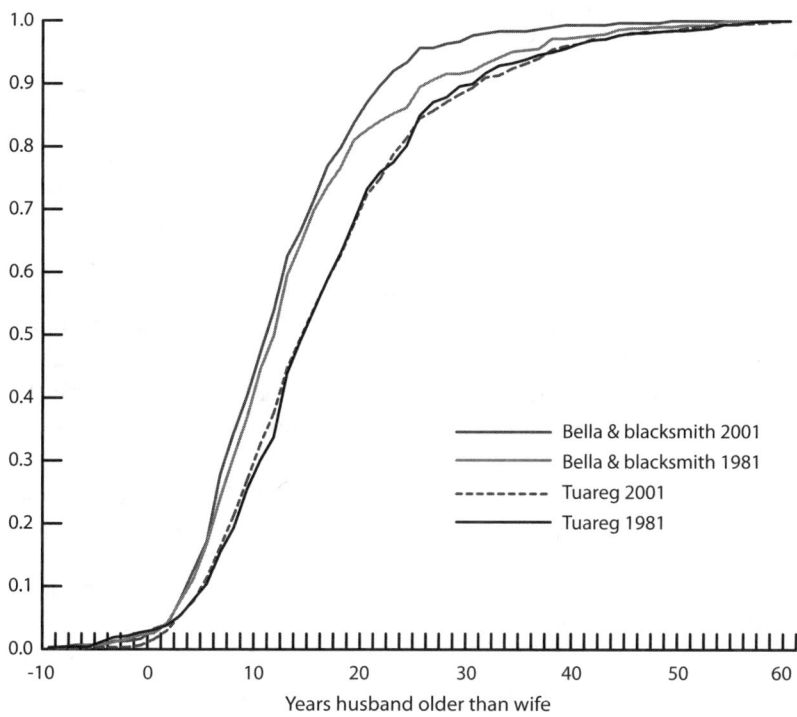

Figure 3.6: *Tamasheq spousal age differences by social class, 1981, 2001*

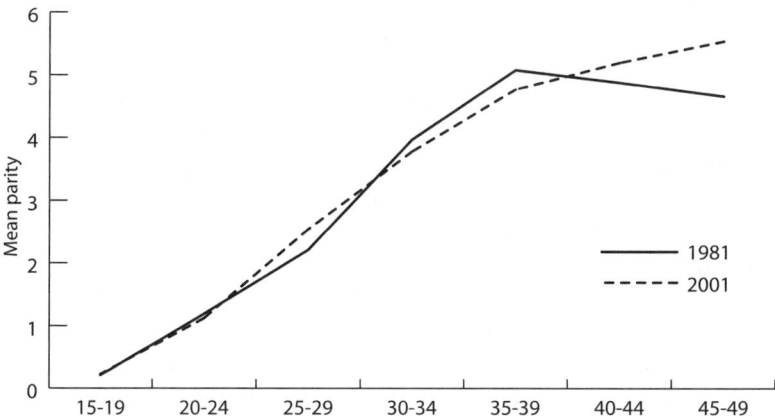

Figure 3.7: *Mean parity by age of Tuareg women 1981, 2001*

Tuareg perceptions of fat and fertility

The Tuareg perceive themselves to be a population with low population growth rate compared to other populations in Mali. These ideas were already present in 1981 (possibly as a result of educated Tuareg reading the French archives and reporting them back to the population (Randall 2009)) and still persist, post-rebellion, having become accentuated by the ethnic nature of the conflict and subsequently by issues relating to decentralisation and relative voting power. They ascribe their low population growth rates partly to monogamy, partly to high levels of spontaneous abortion, and partly to the intrinsic lower fertility of Tuareg women. When they were living in the Delta they also blamed the unhealthy environment (to which, as desert people, they were not adapted) for high mortality.[14]

In 1981 and 1982 they seemed to be a highly demographically aware population. They were conscious of their own population size and dynamics relative to others in Mali. This was likely to be a function of their perceived vulnerability within the country (later reinforced by the 1990s rebellion but stimulated by the response to a previous rebellion in the early 1960s). Many Tuareg did not conceal their disdain for the black African leaders of the country and they were concerned about their own population dynamics because they were a conspicuous minority. This awareness led people (both literate and illiterate Tuareg, men and women) to talk about issues related to reproduction, fertility and mortality. Low fertility was frequently referred to but primarily as an intrinsic characteristic of being Tuareg rather than caused by any form of behaviour and never specifically related to fatness and force-feeding. If anything, by rendering women more beautiful and sexually attractive obesity would be seen to increase fertility through encouraging marriage. High frequency of spontaneous abortions (for which there is no concrete evidence) was occasionally blamed by educated Tuareg on the high levels of consanguinity but never on obesity.

Obesity is disappearing fast in this region and in this Tuareg population. In the refugee camps there was no surplus food for force-feeding girls. Most households had lost their livestock. Although, post repatriation, many continued to live either partially or fully from livestock herding, huge surpluses of milk and butter became a memory of the past. But force-feeding has not just been abandoned simply because of economic crisis. During the rebellion obese women were a major liability needing camels or oxen for

transport and hindering rapid flight. In 2001, many people remarked that force-feeding was no longer appropriate and should be abandoned, and now it only occurs in one or two anomalous camps and families. Fat remains beautiful and most thin women would like to be plumper, but not to the extent of the past.

Conclusion

Throughout the last eighty years low fertility has been a recurrent theme in discussions of Tuareg demographic dynamics. This was unusual within a region where high fertility was otherwise universally prized and where both women and men traditionally acquired much status and power through having many children. Malian Tuareg were different in many other ways too: physically distinct from the black African majority; nomadic cattle herders rather than sedentary farmers; a monogamous Muslim population surrounded by animist and Muslim polygamists. Having women who ideally did no labour and were not only totally unproductive, but were in fact enormous consumers of surplus production, added further to the particularities of Tuareg society relative to the surrounding populations. It was hardly surprising that the French administrators tended to assume that many of these characteristics were linked: with the administrators' concerns both for the growth of the populations under their control and their determination to 'civilise' their subjects, 'low fertility' had, for them it seems, become somewhat of an obsession. So, for the French, low fertility was certainly related both to obesity and the marriage regime.

From the demographic perspective, the French administrators were correct about the important role of marriage in low Tuareg fertility compared to neighbouring populations but they were wrong about which aspect of the marriage regime. Marriages of young girls to older men were not the major problem: monogamy was the issue. Monogamy, coupled with a mean spousal age difference of nearly ten years (in 1981, girls marrying on average around age seventeen and men around age twenty-seven), meant that there were just not enough men for all the women of reproductive age. The fact that some older men also drew on the pool of very young reproductive-aged women will have contributed to easing the problem of inadequate numbers of men: ironically, it was these marriages that the French abhorred. It is true that these cross-generational marriages probably also involved force-fed girls because the fattest girls were those for whose rich and more powerful families desired good affinal links. Hence one can speculate that fertility (via marriage and the subsequent reproduction) and fatness were linked but in a manner that would be more likely to increase fertility for those concerned.

In fact there is no evidence from the demographic data – whether within population comparisons using data from the 1980s or the analysis of change and absence of change between 1981 and 2001 – which suggests that there was any significant impact of fatness and force-feeding on fertility, although we are hampered by a lack of individual level data and the fact that we can only deal with aggregate behaviour. If there were an impact it would have been indirectly through altering marriage behaviour rather than directly.

From the Tuareg perspective low fertility was not considered to be related to fatness, but neither was there a suggestion that force-feeding was undertaken to enhance fertility. They believe that Tuareg women have more miscarriages than other Malian women, although it is unclear where this idea comes from (certainly not from any concrete data) and it may reflect the colonial administrators' ideas written in the archives. From the literature on obesity and fertility in Europe it does seem biologically plausible that extreme obesity could contribute to somewhat increased fetal mortality, as well as increased risks of stillbirths and neonatal mortality. However the highly endogamous marriage pattern probably played a more important contribution: in 2001 over half of marriages were between

Table 3.3: *Different perceptions of the determinants of low Tuareg fertility.*

Colonial Administration	Demographic Analysis: 1970–2000	Tuareg Perceptions
1. Force-feeding and obesity	1. Monogamous marriage regime	1. Miscarriages
2. Marriage patterns (a) precocious marriage of girls to old men (b) high bridewealth (c) frequent divorce	2. (In the Gourma) something else, possible STDs	2. Intrinsic low fertility of Tuareg women
3. Miscarriage because of 'women's condition'		3. Marriage regime

Source: Author

first cousins and less than 20 per cent were between unrelated couples (Randall 2005); such endogamous behaviour goes back many generations, making them a highly inbred population. According to Hussain (1998) – in his study of the impact of consanguinity on spontaneous abortion in Karachi – consanguineous marriages, and more important, multi-generation inbreeding, have a small but significant increased risk of spontaneous abortion.

Overall there is little evidence to suggest that Tuareg fertility was significantly enhanced or reduced by force-feeding and obesity. However there is much more evidence that force-feeding had an impact on overall reproduction and population growth. This was through two principal pathways, the most significant of which was the health and mortality of obese women and their offspring. As noted earlier, estimates of maternal mortality for this population are extremely high, and although much of this high mortality is probably through obstructed delivery, postpartum haemorrhage and post-partum infection, the risks identified to obese women in good health care, developed country contexts will have been far more acute and serious in a context with no access to ante-natal, post-natal or obstetric care. Moreover, even for surviving Tuareg women their offspring were much more likely to die than those of the Bella. In recent years, since the rebellion, Tuareg child mortality has declined precipitously, partly due to immunisation and healthcare received in the refugee camps; partly because of the fact that the population no longer transhumes in the insalubrious Delta and their water supply is infinitely better; but, also, probably because mothers are no longer obese and no longer subcontract much of the childcare to Bella nursemaids (Randall 1984). Children receive better care with greater continuity, and more physical attention. This means that a far greater proportion of children who are born now survive through to adulthood.

The other pathway linking fatness and fertility is 'marriage'. All three perspectives on Tuareg low fertility cite the marriage pattern as a major factor even if they differ in the aspects of marriage which are highlighted. Tuareg marriage is unusual in rural Mali in that it is not primarily about an acquisition of female labour and only partially about an acquisition of reproductive rights. Unlike populations with exogamous lineages where a lineage needs to acquire women from other lineages in order to reproduce, Tuareg men can marry patrilateral, matrilateral or cross cousins. Marriage is not about acquiring women and children for the lineage, but is more concerned with reinforcing particular links and obligations and developing pathways through which resources can be accessed and power can be consolidated. Obese young girls demonstrated the wealth in cattle and Bella of her natal household but their fatness also signalled that this was a traditional girl, brought up to respect traditional values and who should remain largely immobilised in her husband's tent but would be expensive to maintain. The main constraint on Tuareg population fertility is the fact that there is no pre-marital reproduction and that monogamy means a substantial proportion of female reproductive years are spent outside marriage: either unmarried, divorced or widowed. Being Muslim, polygamy is theoretically possible, and attempted by some men who are usually thwarted by the first wife refusing to accept it and returning home demanding a divorce. Most men would not attempt to have more than one wife because wives are so expensive to maintain in food and goods. So obesity and all that went with it also contributed to many aspects of maintaining the marital regime and thus reduced population fertility.

Future prospects

Force-feeding for adolescent girls is unlikely to return for this population, although socially valued fatness and little energy expenditure will probably mean that many wealthy Tuareg women are likely to become overweight as they age. Whether the marriage regime is maintained remains to be seen. Elsewhere, I have shown that it was remarkably robust throughout the rebellion and repatriation (Randall 2005) but the new independence of young women may change things in the future. Having experienced an active and varied social life in the refugee camps, no longer constrained by kilos of flesh, young women are more physically mobile, walking from camp to camp, visiting and travelling; they are becoming more independent, choosing different fashions and refusing marriages that they do not want. They are unlikely to subscribe willingly to polygamy, but divorce is likely to remain acceptable. Hence, one would predict that the marriage regime and therefore the fertility will probably remain stable for the foreseeable future but with an increase in ages at first marriage.

Acknowledgments

The research in 2000 and 2001 was funded by the U.K. Economic and Social Research Council, Grant Number: R000238184 and was undertaken in collaboration with ISFRA, Bamako. Alessandra

Giuffrida undertook the anthropological study and I am grateful for her work and all the insights she has provided. The 1981 to 1982 demographic studies were commissioned and financed by ILCA with further financial support from the Population Council.

Notes

1. The terminology of Bella and Tuareg will be used here as a simplification although there are several different endogamous Tuareg classes. Because of small numbers, data on blacksmiths will usually be combined with the Bella although in terms of the traditional Tamasheq class groups they are 'free', not captives. Most blacksmiths are black African, they were not persecuted during the rebellion, they were not force-fed and blacksmith women have always been economically active.

2. Most people in the Mema left because there was nowhere there to hide. Further north, around Goundam and Timbuktu, some fled but others hid with their animals in the mountains and the desert. The massacres in the North were later – around 1994 – and more people fled then.

3. The Tamasheq term for the ex-slave class is *iklan*, which, although still used by some high status Tamasheq, has pejorative overtones. The Songhay term – Bella – used by many Malians will be used here.

4. The Bella proportion of the population was always much higher in the more southern Tamasheq populations (reaching more than 50 per cent), which included the Gourma and Delta populations surveyed. In the far north, Bella were rare and Tuareg women much more economically and physically active.

5. TFR between five and six compared to over seven for other rural Malian populations.

6. In the wet season, most high status women consumed only milk and milk products. Grain was consumed in the dry season but was not considered proper food.

7. Obviously, there were some exceptions of women who were active and women who invested much time and energy into caring for their children: but they were the exception rather than the rule.

8. This is usually a death sentence for the child unless another woman whose child has died is prepared to breastfeed it. Amongst the 1,300 women interviewed in 2001, only one had been unable to breastfeed because of illness.

9. Ideally, such data would be available from the demographic survey but the age data were so poor that often year of menarche was the only available marker and it was generally assumed that girls were fifteen at menarche and their age was estimated from this point.

10. A representative sample survey undertaken in Mali in 1962 and 1963 excluded the northern regions of Mali. A demographic survey of nomads was undertaken in Niger (1966) using several different sampling approaches: but the data quality is very poor.

11. About 1,300 women were interviewed. Some of the older ones remained overweight but not obese in the way they had been in 1981. Only one young girl aged about twelve was observed who had been force-fed. There was one highly traditional camp that the author was not allowed to enter because the chief was highly religious and strongly opposed to NGO interventions and Europeans. The team collected data in this camp (omitting the questions on contraception) and stated that the women there were much fatter than we had seen elsewhere during the survey.

12. Ideally, one would compare the age specific fertility schedule but the age misreporting in 1981 means that this graph is substantially distorted, whereas cumulated parity by age is less sensitive to age misreporting.

13. The deficit of births for older women in 1981 is probably due to underreporting.

14. To an extent, they were correct since they are likely to have low levels of sickle cell trait, which protects against malaria: the Delta not only had appalling water quality, where people mainly drank standing water, but also very high incidence of malaria. Mortality amongst Tuareg is significantly lower now that they no longer spend the dry season in the Delta.

References

Cedergren, M. 2004. 'Maternal Morbid Obesity and the Risk of Adverse Pregnancy Outcome', *American Journal of Obstetricians and Gynecologists* 103(2): 219–24.

Chambers, R., et al. 1981. *Seasonal Dimensions to Rural Poverty*. London: Pinter.

Donath, S.M. and L. H. Amir. 2000. Does Maternal Obesity Adversely Affect Breastfeeding Initiation and Duration?', *Journal of Paediatric Child Health* 36: 482–86.

Durnin, J. 1987. 'Energy Requirements of Pregnancy: An Integration of the Longitudinal Data from the 5-Country Study', *The Lancet* 14 November: 1131–33.

Ellison, P. 2001. *On Fertile Ground*. Cambridge, MA: Harvard University Press.

Fulton, D.J.R. and S. Randall. 1988. 'Households, Women's Roles and Prestige as Factors Determining Nuptiality and Fertility Differentials in Mali', in J. Caldwell, A. Hill, V. Hull (eds), *Micro Approaches to Demographic Research*. London: Kegan Paul International, pp. 191–211.

Gesink Law, D.C., Maclehose, R.F. and M.P. Longnecker. 2007. 'Obesity and Time to Pregnancy', *Human Reproduction* 22(2): pp. 414–20.

Hilderbrand, K., A.G. Hill, S. Randall and M.-L. van den Eerenbeemt. 1985. 'Child Mortality and the Care of Children in Rural Mali', in A.G. Hill (ed.), *Population Health and Nutrition in the Sahel*. London: Kegan Paul International, pp. 184–207.

Hill, A.G., S. Randall. 1984. 'Différences géographiques et sociales dans la mortalité infantile et juvenile au Mali', *Population* 39(6): 921–46.

Hussain, R. 1998. 'The Role of Consanguinity and Inbreeding as a Determinant of Spontaneous Abortion in Karachi', *Pakistan Annals of Human Genetics* 62: 147–57.

New York Times. 4 July 2007. 'In Mauritania, Seeking to End an Overfed Ideal': http://www.nytimes.com/2007/07/04/world/africa/04mauritania.html?e x=1184817600anden=25dd46149330714bandei=5070

Popenoe, R. 2003. *Feeding Desire: Fatness, Beauty and Sexuality among a Saharan People*. London: Routledge.

Prentice, A. and A. Prentice. 1988. 'Reproduction against the Odds', *New Scientist* 118(1608): 42–49.

———, G. Goldberg and A. Prentice. 1994. 'Body Mass Index and Lactation Performance', *European Journal Clinical Nutrition* 48: S78–89.

Raatikainen, K., N. Heiskanen and S. Heinonen. 2006. 'Transition from Overweight to Obesity Worsens Pregnancy Outcome in a BMI-dependent Manner', *Obesity* 14(1): 165–71.

Randall, S. 1984. 'A Comparative Demographic Study of Three Sahelian Populations: Marriage and Childcare as Intermediate Determinants of Fertility and Mortality', Ph.D. dissertation. University of London: London School of Hygiene and Tropical Medicine, pp. 253–59.

——— 1996. '"Whose Reality?" Local Perceptions of Fertility versus Demographic Analysis', *Population Studies* 50(2): 221–34.

——— 2005. 'Demographic Consequences of Conflict, Forced Migration and Repatriation: A Case Study of Malian Kel Tamasheq', *European Journal of Population* 21(2–3): 291–320.

——— 2009. 'La natalité Touarègue: des représentations coloniales aux réalités post-rébellion. in: Mémoires et démographie', in R. Marcoux (ed.), *Regards croisés au Sud et au Nord*. Quebec: Presses de l'Université Laval, pp. 252–259.

——— and A. Giuffrida. 2006. 'Forced Migration, Sedentarisation and Social Change: Malian Kel Tamasheq', in D. Chatty (ed.), *Handbook on Nomads in the 2first Century*. Leiden: Brill, pp. 431–62.

——— and M.M. Winter. 1985. 'The Reluctant Spouse and the Illegitimate Slave: Marriage, Household Formation and Demographic Behaviour among Malian Kel Tamasheq', in A.G. Hill (ed.), *Population Health and Nutrition in the Sahel*. London: Kegan Paul International, pp. 153–83.

République du Mali. 1987. *Enquête Demographique et Santé, Mali 1987*. Bamako: Direction Nationale de la Statistique et de l'Informatique/ Calverton, MD: Macro International.

——— 1995. *Rencontre Gouvernement-Partenaires sur le Nord-Mali, Annexe*. Bamako.

Swift, J.J. 1977. 'Sahelian Pastoralists: Underdevelopment, Desertification and Famine', *Annual Review of Anthropology* 6: 457–78.

Usha Kiran, T.S., S. Hemmadi, J. Bethel and J. Evans. 2005. 'Outcome of Pregnancy in a Woman with an Increased Body Mass Index', *British Journal of Obstetrics and Gynaecology* 112: 768–72.

Wagenaar-Brouwer, M. 1985. 'Preliminary Findings on the Diet and Nutritional Status of Some Tamasheq and Fulani in Central Mali', in A. Hill (ed.), *Population Health and Nutrition in the Sahel*. London: Kegan Paul International, pp. 226–53.

World Health Organization. 2007. *Maternal Mortality in 2005: Estimates Developed by WHO, UNICEF, UNFPA, and the World Bank*. Geneva: World Health Organization.

Chapter 4

WOMEN OF GREAT WEIGHT:
FATNESS, REPRODUCTION AND GENDER DYNAMICS IN TUAREG SOCIETY

Saskia Walentowitz

Theoretical frame

The fattening of young girls by milk and millet was a common practice amongst Berber and Arabic speaking nomads of the Saharan desert, until the devastating droughts that occurred during the last decades of the twentieth century. If it is now only performed in some parts of Mauritania as well as in a few Arab communities in Niger, female embonpoint still incarnates a beauty ideal that would be labelled as 'pathological obesity' in the West. Whereas fat female bodies reinforce negative judgments about women in Western societies, Saharan force-feeding and fattening practices are regarded as obvious signs of 'female oppression'. Women's magazines, newspapers, websites and television – all media that constantly praise the reverse body extreme of slimness – regularly accuse the 'violence' of fattening as a survival of 'archaic culture', together with an anachronistic practice of 'slavery'. 'Slaves', it is argued, free the fat, hence 'leisureous' women from household cores, thus taking for granted that the latter would otherwise be women's work.

Beyond common sense explanations, the few anthropologists who attempt to explain the practice of fattening focus on its aesthetic values. According to Aline Tauzin (1981, 2001), force-feeding indeed acts as an instrument of 'male domination' that disciplines female sexuality and transforms women into perfect 'objects' of male desire by preventing women's own desire through the oral satisfaction of food. Without making any statement about the possible links between fattening and gender hierarchy in Mauritanian society, nor discussing the relevance of psychoanalytical theory for social anthropology, it seems that the limit of Tauzin's analysis is that it tends to reduce gender relations to sexual relationships between individuals. Rebecca Popenoe develops a different approach in her monograph on fattening among the Arabs of Niger, demonstrating how this beauty ideal is embedded in wider social and cultural meanings in which women play a rather active role through the management of 'all-powerful forces of sex and desiring' (2004: 197). Here, the concept of 'desire' is associated with a range of gendered social dynamics, which are not bound to a separable domain of 'sexuality', but are expressed through the idea of 'appropriate' mutual desire, that is both created and controlled by the beauty of fattening.

In agreement with a necessarily sociological approach to gender relations, while considering in addition that these are not subsumed in the sole relationships between 'women' and 'men', I will explore the practice of fattening among the neighbouring Tuareg nomads by emphasising its meanings with regards to the gendered dynamics of social relations. For doing so, I will follow Annette Weiner's lead towards gender relations understood within complex processes of 'reproduction'[1] to which women and men contribute differently and in various roles at all levels of social and cosmic life (1992, 1995).

In her broad understanding of intricate human and social reproduction, Weiner distances herself from views that bound gender relations to sexuality and procreation in Western societies as a result of historical transformations:

> Throughout the world, human and cultural reproduction are sources for the most powerful legitimating forces of social life. In Western history, however, the economic and political necessity to control private property within the nuclear family resulted in the conflation and marginalization of women's reproduction and sexuality. The focus on the nuclear family to the exclusion of other kin relationships made women's roles as wives critical to this legitimization. The economic priorities continue to form the bedrock of Western social theory and practice. (Weiner 1995: 407–8)

In the context of nomadic societies of the Western Sahara, such methodological and theoretical bias may be even more difficult to identify and overcome in that the relationships between men and women as potential partners are culturally sophisticated, thus apparently confirming the centrality of 'sexuality', while human reproduction, at least among the Tuareg, does not appear to be central in women's lives. Furthermore, the well-documented ethos of gender balance that shapes Tuareg society in cosmology and social practice invites us to increased 'epistemological awareness' (Bourdieu, Chamboredon and Passeron 1968) in order to avoid value judgments which hinder the understanding of fattening as a part of a complex social whole. Tuareg women indeed enjoy an exceptional high status, including not only great respect, but also economical autonomy. They are engaged in extended gift economies based on inalienable women's wealth called 'living milk'.[2] These collectively owned possessions include not only reproducing livestock and trees but also objects of prestige such as swords as well as the 'war drum' (*ettebel*)[3] of chieftainship (Claudot and Hawad 1987; Worley 1991). Among the Azawagh Tuareg, where I collected most of my data, this type of wealth is also known by the Arabic term *el habus* and frequently consists of animals that are donated by important ancestors to a group of siblings and their descendants. Even those animals, which are inherited by sons according to Islamic jurisprudence, are in practice left with sisters. In general, taking animals back from a sister is a shameful act for a man.

All together these 'inalienable possessions' embody the continuity of the society that is however put into question today because of general pauperisation: due to the impacts of colonization followed by political and economical marginalization in post-colonial states, along with ecological crises and, nowadays, new forms of conflict in the context of global competition over uranium and other non-renewable resources found in Tuareg country.

Despite dramatic social change, women continue to be considered as the 'central pillars' of society, as those who maintain social cohesion through the careful management and mediation of various relationships including those between the living and the dead, humans and non-humans, between kinsmen, allies, and cultural others, as well as between members of various social strata that make up the hierarchical but complementary parts of society (Claudot-Hawad 1990). Women are the nexus of sociality, a position that is visible through the ideal of female embonpoint and immobility. The body here is not the point of departure of 'cultural constructions' of 'womanhood', as opposed to 'manhood', but embodies cultural principles and values that shape broader gendered social dynamics.

This is best demonstrated through the analysis of the articulations between reproduction and kinship in Tuareg society. It is useful to refer to Weiner's analytical recognition of women as sisters within her concept of 'sibling intimacy' defined as 'a broad range of culturally reproductive actions, from siblings' social and economic closeness and dependency to latent, disguised, or overt sexual relations' (Weiner 1995: 411). Weiner states that siblingship remains a blind spot in social theory as well as in kinship theory because of the moral strength of the sibling-incest taboo which excludes siblingship from marriage and, consequently, from procreation. According to her, this gap is particularly striking in Lévi-Strauss's alliance theory that reduces women's reproductive roles to wives exchanged by men. The limits of alliance theory, as well as of descent theory, have been discussed by most authors of Tuareg kinship studies (Bernus et al. 1986) who often end up underlining the centrality of siblingship, but without investigating it any further on a theoretical level (Walentowitz, *forthcoming*).

In this chapter, I will try to analyse gender dynamics from a Tuareg point of view by focussing on opposite sex siblingship as a core feature of kinship and reproduction. I shall therefore briefly expand upon my approach to Tuareg kinship, which has proven, like kinship in other Muslim contexts, difficult to analyse, because it combines unilineal, most often matrilineal, descent principles and a strong preference for 'endogamy' with cognatic kinship and 'exogamy'. As soon as one reintegrates siblingship into kinship theory, these apparent analytical contradictions become most coherent. At the same time, the concepts of sibling intimacy and reproduction shed new light on kinship as 'representation' and kinship as 'will' (Bourdieu 2000 (1972)) beyond the much debated dichotomy of structure and agency. At this point, it is important to underline that it is not only a broad understanding of reproduction, but also the strong emphasis placed by Tuareg on genealogical ties, which calls for a critical revision of the articulations between reproduction and kinship.

In this perspective, I will try to show that the analysis of female fattening allows us to make these articulations explicit, even if the Tuareg themselves do not explicitly connect fatness with procreation in the narrow sense of enhancing 'fertility'. Although they consider the fact that extreme corpulence makes pregnancy invisible to be one of the advantages of fattening. My argument is that siblingship is not only central to reproduction but also to procreation itself, in a society where 'procreation' and 'sexuality' are culturally disconnected and reframed in terms of sibling intimacy. Fattening participates in these processes of disconnecting and reframing reproductive forces of women and men as siblings and spouses.

Ethnographical frame

The ethnographic data presented here draws on a long-term study conducted between 1996 and 2000 among the Azawagh Tuareg of Central Niger (Walentowitz 2003). I shared life for more than two years with the Inesleman community of the Kel Eghlal and the Ayttawari Seslem, the two main groups that constitute the religious and political elite of the former Tagaraygarayt confederation. Although Tuareg have been Muslims for several centuries, few studies have been conducted among the 'patrilineal' religious specialists of this highly stratified pastoralist society. In the anthropological literature, the social hierarchy is generally described distinguishing 'noble' warriors (*imajeghan*) and their 'vassals' (*imghad*), religious *Inesleman* pacifists with a status comparable to those of the *imghad*, some religious warrior groups like the Azawagh Inesleman, similar to the 'nobles', the *inadan* craftsmen who also play important roles in ritual and politics, and former 'slaves' (*iklan*) of various social strata according to their roles and integration into the groups of their former masters. Beyond this general description, the local logics of social hierarchy and power relations still remain to be critically studied in detail, beyond Eurocentric preconceptions such as 'feudalism' or Marxist-inspired social theory, and beyond the individual-focussed human-rights approaches to 'slavery'. Indeed, 'slavery' was common to all societies of historical West Africa and was found elsewhere in the continent.[4] Referring here only to the Inesleman's political elite of the Azawagh Valley, I should underline that their kinship system developed in the course of a complex but poorly studied history of changing political institutions and religious reforms dating from the seventeenth century onwards (Norris 1975).

By the end of the nineteenth century, just before the French colonization, the Azawagh Tuareg were, like the Almoravids, organized in a confederation headed by a secular political leader chosen among the Imajeghan warrior elite and a religious political leader recruited among the Inesleman (Walentowitz 1999). The political institution of the latter was designated by the term imamate (*telemumiya* in Tuareg language): the title of imam (*elilam*) was not inherited with reference to tribal affiliation, but offered through election to the 'most knowledgeable, experienced and wise', in a way similar to that of the early Berber imamates of Kharejite obedience. Within this general context, the person who has the abilities to become the imam is a 'person with a lot of kin', within an ego-centred kinship system that valorises a maximum number of genealogical connections with the same agnatic ancestor in a wide web of cognatic kin. Amazingly, this kinship system stresses plural genealogical identity while operating with close-kin marriages as well as with unions among distant kin in order to connect various sibling sets over several generations. It is this apparent paradox of a continuous creation and reinforcement of proximity with simultaneously a maximal extension of kinship bonds that correlates with the ideal of female embonpoint and immobility and its pendant, male thinness and mobility.

Getting fat among the Tuareg

I should begin with a general description of 'fattening' – a procedure no longer existing among the Tuareg – which I learned of from older women and those who in the 1960s and early 1970s were the last generation to be fattened. Among the Tuareg of the Azawagh, who live close to the Arab community of Popenoe's ethnography (cited above), young girls started to be fattened at the age of seven. The initiation of fattening began after a 'woman of renown' (*tamassarhayt*) performed the first hair braiding of the young girl. During this braiding ritual that, unlike fattening, is still carried out nowadays, the elderly woman traces the five-part structure that is characteristic for all female hair arrangements, beginning with a circle at the top of the head. Once this circle is traced, it should never be removed, otherwise the popularity and power of attractiveness (*sarhaw*) inherited by the woman of renown would supposedly vanish at her death. Together with the first braiding that is collectively completed by all women present on that day, the ears of the young girl are pierced with silver earrings, which become larger and heavier over the years of fattening.

It was usually a descendant of *iklan* who fattened the young girl. This woman was supposed to be a 'slave of origin', in other words a slave with a pedigree, whose ancestors had been integrated into the family for several generations. Using a wooden funnel, she forced the girl to absorb large amounts of fresh cow's milk, diluted with some plain water. She also made her swallow handfuls of pounded raw millet *(tedda)*. The Tuareg refer to this custom as well as to the practice of fattening, as 'filling' *(adanay)*. In the beginning, the young girl was force-fed until she got used to it and started to participate routinely by lying down on her back and pulling her lips wide open. If she resisted, the woman shouldn't hesitate to pinch her or to pull back her fingers, sometimes even until they broke. Fattening, as a crucial part of a woman's education, was associated with physical and moral pain and can be indeed considered as an 'extended initiation' (Popenoe 2004: 197).[5] Older women like to talk about their fattening experiences while proudly exhibiting their most appreciated stretch marks and distorted skin folds under their arms.

Usually, the girl ingested fresh milk diluted with water two times a day: in the late morning and in the late afternoon. When asked how much milk they drank per day, many old women like to answer, 'It was so much milk that I was urinating the whole day.' Pounded raw millet was used as a complementary food or as the main food during the dry season when fresh milk was lacking. In any case, the young girl had to avoid taking in milk and millet at the same time. The fattening diet was indeed strictly dissociated in order to prevent stomach sickness *(tagnut)*. It was only later, when the girl had gained much weight, that she started to eat cooked porridge mixed with some fresh milk *(elliwa)* or sour milk *(essesli)*.

Among the Tuareg, uncooked food is considered 'cold' and cooked food regarded as 'hot'. Fresh milk as well as sour milk is 'hot', whereas fresh milk blended with water is 'cold'. Tuareg women are very sensitive to the 'heat' or 'freshness' of food and liquids. They try to keep a balance, although a woman should not eat too many 'hot' foods like butter, meat or salt.[6] 'Cold' fattening foods are said to create a fat, cool, moist, smooth and light-skinned body that corresponds to the ideal female body. Men must be just the opposite: slim, 'hot' and 'dry'. They are not allowed to eat *tedda*,[7] which is a woman's food *par excellence*, and are subject to mockery, both by women and fellow males, if they start to show a tummy.

Beside the fact that men must not eat uncooked millet, there is also a major difference between the way men and women consume porridge. Whereas women blend milk and porridge in order to swallow it in large amounts, men are supposed to take some thickly cooked millet porridge *(esshink)* with a wooden spoon and dip it into some fresh or sour milk contained in a different bowl. I was unaware of this detail until I ate porridge this way with a young woman who observed me for a while and suddenly exclaimed: 'Oh, you also prefer to eat like a man! So do I!' This young lady had not been fattened, but until today, girls are taught to ingest large amounts of semi-liquid milky porridge by swallowing without tasting it.

Fattening in a gender-balanced society

Tuareg consider milk as a female-gendered food, whereas millet is regarded as a male-gendered food. 'Cold' milk and 'hot' millet are seen as perfect complementary ingredients[8] and build the basis of the nomad's daily meals. The fact that it is only women who are allowed to blend the two after a long period of dissociated fattening diet is a significant detail that leads us directly to the realm of reproduction. The Tuareg, as I stated above, do however not link 'fertility' with this practice explicitly. Indeed, when asked the question why they fatten their womenfolk, most Tuareg, women and men, answer that fat is beautiful and that fattening transforms a girl into a woman, physically, emotionally and spiritually. Beauty is highly praised in oral poetry depicting portraits of women with endless black braids and light soft skin, shiny teeth, dark eyes and a pleasant body with rolls of fat falling down to the knees (Drouin 1984). At the same time, fat and beauty do not by themselves make a great woman: intelligence and knowledge are of equal importance in this society where maternity is not a precondition of 'true' womanhood (Claudot 1984), and where male infertility is not only recognized but seen as incurable, unlike female childlessness (Walentowitz 2004a).

Although the practice as well as the aesthetic values of fattening are very similar among the Arabs and the Tuareg of Niger, it is, indeed, the central position that women enjoy among the latter that challenges the analysis:

> Because I am suggesting that Moor fattening is deeply embedded in structures that are particular to Moor life, the relationship of Moor fattening to Tuareg fattening deserves some comment. Tuareg culture and Moor culture share Islam, a history of slave holding, and a dependence on herding and, to a lesser degree, trade. [...] Tuareg culture differs vastly from Moor culture, however, in the relative equality between the sexes that characterizes it, and in the relative freedom of women. While the two neighbouring cultures shared the sexualized aesthetic that surrounds female fatness, as well as some of the foundational structures that undergird the practice, the crucial

differences in constructions of maleness, femaleness, and gender relations between the two societies make it impossible to completely equate the two traditions of fattening. (Popenoe 2004: 39)

I fully agree with the author's views, including the idea that fattening is a 'physical expression of central cultural values' (Popenoe 2004: 50) within a society where women's freedom from household and herding chores does not mean that women are not a driving force of society: 'women's work' is 'stomach work' and has much to do with 'feeding desire' as suggested by the title of Popenoe's book. Female fatness is not simply a sign of wealth, but women are the wealth:

> Out of milk and millet they create sensual beauty, well-being, and reproducibility, turning men's production into something uniquely Arab, while concretizing and encapsulating in forceful idiom that which is of paramount value and meaning. In this work that is not work, women manage the all-powerful forces of sex and desiring, eloquently contributing to the vitality of their world. (Popenoe 2004: 197)

The difference between Arab and Tuareg grammars of gender(ed) dynamics becomes clear when one examines the different meanings and values associated with the themes of desire, attractiveness and wealth that are equally central to the understanding of female fattening among the Tuareg. In Tuareg society, where women are economically independent and where women's wealth is central for reproduction in Weiner's sense, it is not 'men' as a homogeneous category that invest their wealth into 'women', but grandmothers, mothers and aunts who allocate parts of the inalienable female possessions of their group to their daughters, sisters and nieces, while maternal uncles, brothers, and also fathers and husbands invest their possessions in women's wealth as well.

Girls are fattened with 'living milk' and other animals with a pedigree (*elessel*, 'origin').[9] Milking animals of the girl's personal livestock that she starts to constitute right after birth[10] may also serve this purpose. The frequently expressed idea that fattening is a sign of wealth here earns a specific meaning since Tuareg women literally embody inalienable possessions, as well as the continuity and independence of their kin group that this wealth both symbolises and guarantees materially. This embodied continuity is not put into question by marriage. On the contrary, when a woman marries, she takes her animals along to her husband's camp[11] and continues to

feed on her own milk. In former times, female slaves, especially
those women who used to pound the grain for her in a proper way,
accompanied her. The idea of food continuity, including an
invariable, familiar way of pounding millet, is central in perceptions
about woman's health. Tuareg women fear a specific illness called
anaghu, which provokes headaches, allergies and general unwellness
owing to changes in feeding habits. In order to avoid *anaghu*, women
always take their favourite milking animals as well as millet with
them. Some even bring their own water wherever they go. *Anaghu*
is a canonical female illness that correlates with the Tuareg's
emphasis on social continuity and female economical autonomy. A
woman is, and remains, bound to her group of origin, which she
represents and extends wherever she goes.

According to the Tuareg ethos of honour (*asshak*) – which above
all refers to the honour of men towards women – regardless of social
position or kinship proximity, men as brothers must help to sustain
the wealth of their mothers and sisters,[12] and, at the same time, they
must cover the needs of their spouses and children, including all
costs that concern the household and beyond. A woman may
contribute to these costs, especially when it comes to hospitality of
important guests, but it would be shameful for a man to take, for
example, a sacrificial animal from his wife's herd, if he has enough
animals himself or without asking her permission. Weiner's
distinction of women's roles as spouses and as siblings is crucial here
for an analysis of Tuareg society, as it is for many other societies, not
necessarily matrilineal:

> Regardless of how descent is reckoned, the roles of both, spouses and
> siblings reveal the possibilities for, as well as the limitation of the scale
> of authority that is possible in a social system. Frequently, sibling
> intimacy creates tensions in the roles that women and men play as
> spouses. These same tensions appear in the larger kinship and political
> domain, where the need to regenerate the social identities of one's
> kin group stands in opposition to the demands of others outside the
> group. These tensions skew gender relations in particular ways, in
> some cases bringing women into political prominence, in other
> situations securing only male hegemony. But since these tensions are
> located in the reproductive domains of spouses and siblings, they
> reveal the problems that surround sexuality and human and cultural
> reproduction. Symbolically and materially, women and reproduction
> – seen as an integral part of social and political action – must be
> accounted for in social theory. (Weiner 1995: 419–20)

The practice of fattening offers a privileged entry point to the analysis of the competitive, but complementary reproductive roles of siblings and spouses. In the following section, I will make explicit these roles by studying first the theme of sexualised attractiveness of fattened women as (potential) spouses. I will then contextualise this theme within the dynamics of kinship and alliance based on opposite sex siblingship and, finally, analyse the links between fattening, siblingship and procreation, namely through the study of a woman's postpartum diet that is in fact a re-actualisation of a young girls fattening regime.

Fattening and female attractiveness

Fattening seems to be a very old practice among the Saharan Berbers. Ibn Battuta reported in the mid-fourteenth century that:

> the women of the Bardama are the most perfect in beauty, the most extraordinary in their exterior, of a whiteness without admixture, and of a heavy corpulence. I have not seen women as fat in any country. Their food consists of cow's milk and ground sorghum, which they drink morning and evening, mixed with water and uncooked. He who wants to marry one of them must live with them near their own territory and not go beyond Gao and Walata. (Quoted in Popenoe 2004: 34)

The similarity with Tuareg fattening is striking, but the most significant sentence of the above quotation is the last: a man may not marry a woman and take her far away from her people. Although Tuareg are traditionally nomads for whom mobility is also the driving force of social and cosmic life (Claudot-Hawad 1994), women are associated with immobility that is not merely a consequence of fattening but is actively created by it. Whereas men do travel long distances, formerly for caravan trades and warfare, women do not usually leave their territories of reference. In this sense, women do not travel, but only move (Claudot-Hawad 2002; Walentowitz 2002). Female immobility must not, however, be misunderstood as a stereotype of passive and secluded women, while men have the freedom to discover the outside world. Women are, on the contrary, the fixed stars of life. In Tuareg society, there is a reverse polarisation of 'public' and 'domestic' spaces:[13] the fact that women are bound to the interior is not regarded as seclusion, nor does it restrain women from long-distance visits. Men are allowed to enter a woman's tent only if they are entirely veiled, be it as admirers during nightly visits, or as husbands living in their spouse's tent for the time of marriage. Tuareg women also wear headscarves, but unlike men who must be veiled in public space as well as in the presence of any women including those who are close, unmarriageable kin, they must not cover their faces when going outside.[14]

The nomadic tent belongs to women only in the majority of groups and represents a sacred, inviolable space, indeed the indispensable shelter, without which no physical or social life is possible (Claudot-Hawad 1994). A woman's reputation depends on her ability to sustain a 'tent' materially and socially. A woman of great weight is a woman who is able to create large social, economic and political networks, namely through great generosity and outstanding hospitality (Claudot-Hawad 2000a; Walentowitz 2006). The years of fattening correspond to a woman's education during which she literally incorporates the Tuareg system of values. Women learn their role of being a 'central pillar' of society, a role which requires not only knowledge and skills related to sociality at large, but also physical and emotional abilities of 'self-control' (*iduf n imam*, literally, 'the maintaining of souls') and 'compassion' (*tahanint*). Fattening plays a crucial role in the acquiring of self-control through the ingestion of 'cooling' foods, and, of course, the strength needed to endure the practice itself. It is not about controlling a woman's emotions, but about intensifying her feelings and enhancing her ability to perceive these feelings consciously.[15] Milk plays a specific role in this process in that it is linked to the divine breath as well as the faculty of perception. It is also connected with mind and intelligence, and since fattened women drink much more milk than men, the former are regarded as more intelligent. The following story provides us with an illustration:

A father on his deathbed gives his son a stone mill that is cut into two pieces. He advises his son to marry only the woman able to sew them together. After his father's death, the son goes from camp to camp in the hope to find a woman able to perform this impossible task. Each time he reaches a campsite, he only encounters laughter and incomprehension. Just before losing hope, he tries his luck in a last camp far away, accompanied by his blacksmith. Again, the women he meets just think he is insane. On his way back, he crosses the path of a young girl in the middle of a fattening session. She calls the blacksmith and asks to see the mill. As soon as she set her eyes on it, her mouth plenty of *tedda* millet, she spills out some of the sand that is always found in pounded grain. She hands the sandstones to the smith and says: 'Please give these sandstones to your master and ask him to transform them into some metallic pieces [used to repair wooden porridge bowls] as well as into some nails. Bring them to me and I'll repair your stone mill. If you do not succeed, I will return the stone mill to you.' The young man followed the conversation and knew that he had just met his future wife.

After the wedding, the story continues until the couple arrives in another territory where the chief falls in love with the woman and plans to kill her husband while asking him to travel somewhere with his men. The wife realizes the strategy and asks the chief to give her some time in order to prepare her husband's luggage. She provides him with just about everything a person might eventually need during a long trip, whilst advising her husband to be generous with everybody and to help whenever there is need. The man follows his wife's advice and in the end, nobody wants to kill him. Instead, they come back and kill the chief who is soon replaced by the woman's husband.[16]

What this story tells us is that, in Tuareg society, it is the woman who makes a man, not the man who makes a woman. 'A woman is a man's trouser belt,' says the proverb. If it were the other way round, the woman of the story could simply have become the wife of the first chief. Fattening provides the woman with intelligence, spirit and all the values that are at last condensed in the central concept of honour (*asshak*). The increasing size of body is proportional to the acquiring of these abilities allowing a woman to attract the outside to the inside world of her 'tent'.[17] Female immobility thus cannot be ethnocentrically equated with 'passivity'; on the contrary, it is a sign of a woman's capacities to actively sustain social continuity and to improve life. It is this social power of attractiveness that is expressed through a woman's sexualised beauty and sensual power of attractiveness.

Female power of attractiveness is ritualised in rites of passage, including the ritual during which a young woman receives her first headscarf. After several years of fattening and education, and imperatively before her first menses among the Azawagh Tuareg, a young girl receives a headscarf during a special veiling ceremony from the hands of the same woman of renown who performed her first hair braiding (Walentowitz 2002). From this moment on, she is explicitly authorised and expected to participate in lyrical events and receive admirers in the intimacy of her mother's tent. The more admirers she has, the greater is her reputation called *sarhaw*, a term that means 'popularity', 'renown' and 'power of attractiveness' in both, its social and its sensual acceptation. An older woman with great *sarhaw* might offer the girl a pearl of her golden necklace or even a piece of her underclothing in order to increase the maiden's attractiveness. The latter is easily measurable in an interesting way. Indeed, the men who visit her during the night should leave some tobacco under her pillow. In the morning, descendants of former *iklan* and artisans, who craft a woman's reputation, come and ask the woman for some tobacco. Its availability is thus a good indicator for her attractiveness to men, and underlines the importance of generosity and redistribution for a woman's prestige. Tobacco is appreciated by everyone, and so is the woman who offers it to her people and who, in turn, will share it with others.

The institution of nightly visits, often described as evidence for the Tuareg's 'sexual freedom', contributes in fact to a woman's education as a 'woman of great weight'. The centrality of fattening in this process appears in stories of a nightly visitor who inadvertently flings his javelin into a young woman's extended thigh instead of the ground. In order to avoid the dishonour of the young man, the woman doesn't say a word and engages in a long conversation thanks to the patience she acquired through fattening. In its codified form, these visits are comparable to hospitality rituals, that later in her life, a woman is supposed to lead to perfection. Hospitality is the core of sociality among nomads and symbolizes the real wealth of a person: the wealth of people.[18]

The fact that female beauty is not only meant to please men, but is embedded in a wider concept of 'attractiveness', is also evident in hospitality rituals among women who are most careful with their appearance when visiting other women. Female hospitality differs widely from hospitality offered to men. Whereas a man travels without taking many things along, a woman takes all she needs with her and exhibits her economic independence with great ostentation. Everybody who holds on to his or her reputation of being an honourable person with a great sense of hospitality should satisfy or, better, anticipate any desire of a female visitor who, in turn will immediately make it clear that she doesn't need it by redistributing food and things to those who accompany her.

During life-cycle rituals and receptions, especially during naming ceremonies and weddings, women are competing in beauty, conversation and pride. Their body sizes are at the centre of comments and mockery, more or less expressed in a mode of joking relationships. Women measure their bodies with their headscarves and even humiliate the 'smallest' woman by pushing into her part of porridge, a bone from the animal sacrificed in the name of the child. Among the Kel Fadey of eastern Niger, women even engage in wrestling during naming ceremonies (Worley 1992). These competitions among women confirm their centrality in social life as daughters and sisters who act as most respectable wives in order to keep up and extend the reputation of their group of origin. In the next section, I will further contextualise the theme of attractiveness through a brief description of the gendered dynamics of kinship and alliance. In a society where everything tends to show that a woman is the centre and therefore doesn't 'move', it should not be surprising that women are not 'exchanged' by their brothers through marriage. Tuareg operate with a different logic of alliance and descent, which can only be understood when taking into account siblingship as well as the diachronic dimension of kinship.

Siblingship, kinship, and marriage among the Tuareg

At the beginning of this chapter I mentioned the analytical difficulties of kinship and marriage in contexts where Islam is socially recognized, as the combinations of 'endogamy' and 'exogamy' as well as unilineal descent principles with cognatic kin reckoning resist classical descent and alliance theories. The importance of genealogy in Muslim contexts as well as the elaborate juridical apparatus of concepts related to kinship also makes it difficult to analyse kinship systems solely in terms of local strategies and practices (Bourdieu 1972). Concerning Tuareg kinship, the situation is further complicated by the fact that the normative preference for marriage with the matrilateral cross-cousin that is encountered in many Tuareg groups long misled anthropologists towards Lévi-Strauss's theory of generalized exchange (see Bernus et al. 1986). Pierre Bonte (2000a) has shown that Tuareg kinship is more adequately analysed through comparison to kinship and alliance in other Muslim contexts. According to Bonte, 'Arab' and Tuareg kinship systems are based on gender bifurcation, indeed the 'distinction of the sexes' as the organizing principle of relationship terminologies as well as of the opposition between maternal and paternal kinship units respectively called the 'belly' and the 'back'. The asymmetrical opposition between the 'feminine' and the 'masculine' further appears in Bonte's analysis at the centre of related ritual systems including naming ceremonies and weddings, and is identified as the core theme of perceptions about the transmission of vital fluids. According to Bonte, the significant difference between Arab and Tuareg kinship is that the first emphasises the pre-eminence of the masculine, although without denying the complementary role of the feminine, whereas the latter recognises a balance between the two.

Without developing Bonte's innovating approach in detail nor discussing its implications for a comparative anthropology of Arab and Tuareg kinship systems, I will here formulate the hypothesis that these systems are not based on an abstract opposition between the 'feminine' and the 'masculine', but rather on same-sex siblingship in most kinship systems in Muslim contexts, and on opposite-sex siblingship among the Tuareg (see Conte and Walentowitz, 2009). In all cases, sibling intimacy shapes kinship and marriage systems on 'categorial, jural and behavioural levels' (Barnard and Good 1984) including gender dynamics in ritual and social life. In the frame of this article, it is not possible further to discuss this hypothesis, although I will provide the reader with some ethnographic evidence concerning the centrality of siblingship through a brief description of kinship amongst the Azawagh Tuareg.

Despite emic descent constructs such as the 'belly' and the 'back' which anthropologists have long analysed as the 'matriline' and the 'patriline', Tuareg kinship is better described as a 'cognatic' kinship system in which male and female genealogical mediations are equally important. Amongst the Azawagh Inesleman who claim their origins from male ancestors, group affiliation is not simply established through patrilineal affiliation, but through the accumulation of cognatic kinship ties connecting Ego ideally to the whole sibling set, indeed all children of a paternal apical ancestor, via other sibling sets at various generational levels. In other words, direct ascending kinship lines conflate with collateral descent lines.

Furthermore, in the frequent cases of marriage between members of the two most prestigious religious groups of the Azawagh Tuareg, the Kel Eghlal and the Ayttawari Seslem, kinship ties connecting Ego with both his/her paternal and maternal ancestors are equally valued. From a Tuareg point of view, the more a person concentrates kinship bonds to both ancestors, the higher his/her social status in terms of genealogical connectedness (Walentowitz, 2004b).

For this kind of cognatic kinship system based on opposite-sex siblingship, no classical concept of kinship theory is adequate. The system operates with a preference for close-kin marriage, regardless of the categorial status of cousins with regard to the relationship terminology, which nevertheless distinguishes matrilateral from patrilateral parallel and cross cousins. Simultaneously, it requires marital unions with distant kin or non-kin which are valued as well. Indeed, a person who concentrates not only kinship lines with the apical ancestors of the Kel Eghlal and the Ayttawari Seslem, but also with ascendants belonging to various other groups, enjoys an outstanding social position. Here, each union reshapes the boundaries of ego-centred groups of 'descent' born out of the sharing of genealogical connections notably through 'marriage by permutation' or 'exchange' of siblings among close-kin as well as distant-kin (see Conte and Walentowitz, 2009). Women and men play different but complementary roles in order to guarantee transgenerational continuity. Indeed, women may only marry men with equal or higher genealogical positions in terms of their respective numbers of collateral descent lines connecting them to the same ancestor(s). In short, female hypogamy is not encouraged, whereas male hypogamy is not only possible, but structurally required.

The analysis of this kind of kinship system – here far too incompletely summarised – clearly calls for a diachronic perspective that is also crucial in Weiner's approach of reproduction. Marriage politics are part of a wider logic of attracting 'others' who are part of the same whole.[19] Here, there is no such thing as predetermined 'endogamous' 'descent groups', only structurally interconnected kinship networks which are created and recreated in the process of what I term elsewhere the autopoïetic dynamics of kinship (Walentowitz, in press).

In fine, women marry men with genealogical status equivalent to those of their brothers. Thus, the clothing habits of the women belonging to the religious nobility of the Azawagh Inesleman allow us to explore further the social meanings of female attractiveness in terms of kinship. Unlike other Tuareg women, these women hide their faces vis-à-vis men by pulling a piece of their headscarves as a temporary screen. Even more strikingly, they dissimulate their bodies by wrapping themselves in a woven mat. Up to three women are enclosed in one mat that was held up in former times by two slave women. One must imagine a corpulent woman slowly moving with this garment: it looks like a walking tent. In a similar way, when women of a same camp leave for instance for a newborn's naming ceremony, they gather and take large pieces of cloth, surrounding themselves and walking slowly in a compact mass towards the tent or, nowadays, the house where the ceremony takes place. Echoing Annette Weiner's expression of 'keeping-while-giving' (1992) referring to a forgotten dimension of Trobriand exchange, one could describe this phenomenon as 'leaving-while-staying': women do not go out of their houses but they take their homes with them. Women never leave; they stay even when they move elsewhere. And since they take their own mat along, they do not need to sit down on a mat of their host, but remain in their own territory, while extending it in the same movement.

The articulation between this custom and kinship becomes evident in the fact that women do not dissimulate their faces in the presence of any men, but only in front of potential marital partners. In other words, the veiling of a woman's face does not put men at a distance, but on the contrary, it confirms their relative genealogical position and status and thus, points to them as potential spouses and possible candidates for nightly visits.[20] Here again, a woman's headscarf shows its close association with female power of attractiveness, which gains a precise meaning with reference to kinship. Female embonpoint and immobility constructed through years of fattening stands for an ideal web of kinship and alliance that never reduces, but only increases in size by pulling shared kin towards one's own side. The mat in which moving women enclose themselves symbolises this dense web of kinship that is explicitly compared to a 'fine woven mat'.[21] Furthermore, the mat as well as the windscreen (*shitek*) that protects the tent and/or woman's bed from the outside is a symbol for the nomadic territory.

Fattening, procreation and sibling intimacy

In the last part of this paper, I will complete my analysis of fattening with regard to procreation. As stated earlier, procreation is reframed within reproduction by being culturally disconnected from sexuality and conjugality, namely through women's rites of passage and other rituals that frame the relationship between men and women as (potential) spouses. As 'natural' events, anthropologists often conflate sexuality and procreation into a single analytical entity that has only recently been put into question through their material disconnection by the new reproductive technologies. Tuareg ethnography teaches us that there is no need for technology in order to separate these two domains.

In Tuareg society, married people should show mutual respect at all levels of marital life, including sexuality. In no way is sexuality meant for the satisfaction of men alone. This is already demonstrated by the fact that a child is a result of the encounter of male and female semen, both generated in the back out of spinal marrow and liberated after climax. The birth of a girl is a sign that the woman's orgasm happened first and/or that her semen is stronger than that of the man. Ideally, women give birth to children of both sexes who are equally welcomed by society, and equally important in kinship. Fattening and, later, the maintenance of a fat, attractive and healthy body[22] is linked to 'fertility' in the sense that it stands for the women's ability to enlarge her group's social weight, including by giving birth to children. Nevertheless, a childless woman is not stigmatised, while it is dishonouring for a man to divorce a woman because she is barren. A woman might, on the contrary, separate from a man who does not make her pregnant. Furthermore, a woman should not be bothered with too many pregnancies; a birth spacing of three years should be respected, and for that, a man must accept sexual abstinence as well as 'contraceptive' intercourse practices like *coitus intra crura*.[23] If a woman becomes pregnant again too soon, it is her husband who is blamed for his lack of self-control.

These ethnographic details demonstrate that a woman is not an 'object' of male desire, nor is her main role the birthing of children. Tuareg conceptions around menstrual blood are also instructive in this matter. Some anthropologists, who have discussed fattening (Bernus 1998), state that one of the primary goals of fattening is to hasten puberty, menstruation and fertility, and thus, marriage and procreation. From the point of view of Tuareg culture, this statement does not make sense, since menstruation is seen as a bodily process translating a woman's culturally achieved ability to master her emotions (Figuereido-Biton 2001). According to the Tuareg of Northern Mali, menstruation allows a woman to evacuate an excess of 'heat' caused by consciously refrained emotions such as anger, sadness or passion. By contrast to men, women have the advantage of being able to 'cool' their bodies through regular elimination of blood. The menstruation period is regarded as an opportunity for a woman to retire and regenerate her body, soul and mind. The Tuareg of Mali even call menstruation *temezgedda*, meaning the 'mosque' and transmitting the idea that a menstruating woman is in a state of sanctum. Similarly, the Azawagh Tuareg call menstruation by the name *alghadat* meaning 'culture' or 'tradition'. Menstruation is thus not a 'natural' process passively endured by woman and, consequently, fattening cannot 'hasten' puberty, but it contributes to the creation of puberty as a sign of social adulthood. This is also the reason why a young woman receives her headscarf before she first menstruates. Moreover, menstrual blood is seen as female semen transformed into blood in the absence of conception. In this sense, fattening is linked to fertility, because it creates a woman's biosocial ability to procreate.

The active, if not predominant, role that fattened women play in the generation of social life, is echoed in Tuareg perceptions of the womb. Indeed, rather than a passive 'recipient', a woman's uterus is regarded as a sacred organ called 'the tent of the child'. It is compared to a cold and moist goatskin waterbag associated with protective and regenerative power (Walentowitz 2004a). Among the Azawagh Tuareg, a preterm newborn is bathed in water containing the *aggar* fruit of an acacia tree, used for the tanning of animal skins and the making of goatskin bags. It is immersed daily in this water for a period as long as it would normally have stayed in his mother's womb.[24] This generative power of *aggar* water is also found in the Tuareg proverb 'this person emerged from *aggar* water', meaning that this person looks well and has recovered from a difficult period or a longer disease.

The theme of female autonomy connected with her generative power further appears through the rituals that mark a woman's life-cycle as well as those which precede each of her journeys back to her husband's camp. Whenever a woman is sick or in an advanced stage of pregnancy, or if her husband is absent for a longer period, she returns to the camp of her parents or of a brother and takes her tent as well as her animals, domestics and belongings with her. In the absence of his wife, a man simply falls back into the position of an unmarried man and clandestine lover. Because of a strong avoidance relationship between him and his in-laws, especially his mother-in-law, a man can only visit his wife during the night, but must remain unseen and leave before the sun rises.

As soon as the woman is willing to return to her husband's camp, the latter has to send her plentiful gifts, cloth, headscarves, perfume, millet, tobacco, tea, sugar etc., in short, all the gifts a man must send prior to the ceremony that precedes the bride's first transfer to her husband's camp. During this ceremony, the woman gets her hair braided exactly in the same way her hair was braided when she received her first headscarf, and later, during her wedding ceremony. She also puts henna on her feet and hands as well as black protective makeup on her eyes, she wears golden jewellery borrowed from women with great power of attractiveness, she receives new amulets, puts on new cloth, sandals etc., as if she were to marry her husband again. This ritual is called *tadwit*, indeed, a synonym for marriage (*azalaf*). Once a woman has been fattened and veiled, all female rites of passage that anticipate a new cohabitation with the spouse are shaped by the pattern just described.

This is also the case when a woman gives birth and goes back to her mother's place, even if it is her seventh child. A woman must not be seen to be weak in front of her allies. Only at home can she get proper care, because she is not in an official position of representation. Birth takes place within the woman's tent or the tent of her mother. After birth, the new mother will stay completely immobile during two months of the postpartum period. The interesting point here is that during this period, she is fattened again. The only difference with the initial fattening diet is that the mother does not eat 'cold' foods like uncooked millet. Because of the loss of blood, she has to ingest large amounts of 'hot' foods, namely cooked porridge and undiluted fresh milk. Again, she does not mix milk with millet, and she does not mix female and male foods, but dissociates them. The meaning of this dissociation becomes quite clear through a tiny but significant detail; the woman does not add any salt to her diet. In addition, she also takes some cooked meat juice, also unsalted, which assimilates the juice to uncooked blood. In Berber culture, food without salt as well as blood are typically the food of the spirits,[25] whereas salt is also equated with male semen. The postpartum refattening diet strongly disconnects a woman from her spouse and ties her to a brother (or, if she has no brother, to her maternal uncle or a classificatory brother) who is providing the animals for the meat she eats during the whole postpartum period.

This diet is also meant to assure plenty of breastmilk that completes the child's body and person after birth. In the womb, a child only feeds on female blood that is transformed female semen. While disconnecting with the father of the child, a woman connects with her brother through an alliance with male spirits by the consumption of unsalted food. This ritual strongly echoes Tuareg foundation myths where an initial couple of a brother and a sister give birth to the society without marrying each other, but with the help of a bush spirit who impregnates the woman. In some versions, the '*genitor*' of the child is not a spirit, but an outside stranger, indeed an ally. Finally, in this context, it is also interesting to note that, during the whole postpartum period, a woman stays in the northern part of the tent that is a place connected to the spirits. This detail often puzzles anthropologists because it seems to contradict the idea that a new mother and her newborn are particularly exposed to the dangers of spirits. This is to forget that spirits also have a gender and that those spirits that a new mother and her newborn should fear and be protected from [with metallic knifes] during this delicate period of transition, are female, not male spirits.

The closeness between a woman and her sibling during the event of birth is also striking in the Tuareg naming ceremony. On the seventh day after birth, when the newborn baby gets a name, the female relatives of the mother perform a series of birth rituals that imitate each and every single step of gestation and embryogenesis, in the absence of any man.[26] While the women ritually fashion the child's body, soul and mind, beginning with a ritual bath imitating conception, the maternal uncle of the newborn sacrifices a sheep. Whereas it is the father who performs the naming sacrifice in most Muslim contexts, here the father only provides the animal that represents the child. While cutting the animal's throat, the maternal uncle pronounces the newborn's name, hence recognising the child of his sister, before it can be recognised as a child of its father: to put it the other way round, the mother legitimates the father. Indeed, the mother returns to her spouse's camp as a bride, bearing a child in her arms.

Conclusion

Here we touch onto the very foundations of Tuareg 'sibling intimacy', which operates on the verge of sibling procreation, as shown in the fact that illegitimate children [conceived out of marriage] have no right to a naming ceremony. The reason for this is less a moral or religious condemnation of 'children without fathers', than the confusing idea that a child recognised by its maternal uncle in the absence of a father would be much too obviously a child of a woman's brother alone. Thus, fattening is the very process through which 'sibling intimacy' is both constructed and dissimulated. When a Tuareg woman refattens after childbirth, it is said that her belly should be as big after delivery as it was at the term of nine months of pregnancy. Thus, it is as if birth is a non-event, as if after all Tuareg women procreate alone. In real life, the art of women of great weight is to channel the reproductive forces of their siblings and spouses to transform a desert into a vital space, without ever having to leave the shadows of their tents.

Notes

1. In this article, I use the term 'reproduction' when referring to Weiner's broad definition including social reproduction, and the term 'procreation' when addressing human reproduction.
2. It is also called 'milk of the *ebawel* or *abatol*'. The latter terms designate a cavity and imply protection; by extension it means 'female ancestor', 'uterine kinship' or 'matriline' (Claudot and Hawad 1987).
3. The 'war drum' is a large version of the wooden bowl (*tazawat*) representing the origins of humanity and pluralistic power in Tuareg mythology. In daily life, this bowl is used by women for the distribution of meals and hospitality rites (Walentowitz 2006).
4. For a holistic approach of Tuareg 'slavery', see Claudot-Hawad (2000b). The author adopts a sociological and diachronic perspective in order to study the integration processes of *iklan* and the transformation of their collective identities over time. See also the monograph dedicated to the dynamics of *iklan* identity of Burkina-Faso by Bouman (2003). The author underlines that *iklan* cannot be merged into a single static category of former 'slaves'.
5. Without legitimating any form of violence in the name of 'culture', I would like to underline that anthropologists easily recognise the valorising effects of painful extended initiations performed on men, for example among the Baruya of Papua New Guinea (Godelier 1982), whereas comparable treatment of women is spontaneously analysed as 'male domination'.
6. See also Figuereido-Biton (2001).
7. The Azawagh Tuareg tell the story of a man who went to a market in a distant Hausa land. He tasted some *tedda*-like millet called *do* in Hausa and decided to buy some of it for his daughters. Unfortunately, he did not remember the Hausa name well and bought some dark and bad smelling spices called *dodawa* instead. Back home, he discovered his error and claimed: 'Mighty lord, this is not what I have eaten at the market!', revealing that he had shamelessly tasted some food that is strictly reserved to women (story collected in Tuareg by the author, Niamey (Niger), February 2003).
8. Tuareg do not appreciate millet porridge with vegetables or other cooked sauces that are usually served among the Hausa and Zarma.
9. Azawagh Tuareg memorize genealogies for camels as well as horses. Milk is a sacred substance, invested by the social relationships into which the milking animals are tied in. Soon after birth, the gum of the newborn child is rubbed with a date chewed by a religious erudite man while it is breastfed for the first time by a woman of renown. If no such woman is available, the child is breastfed by a 'slave of origin' whose genealogical position enables her to replace this woman. If no suitable woman is available, the newborn is fed on camel's milk from an animal

with a pedigree. Even if a camel stands next to the tent of birth, one might get a specific camel from afar in order to perform the ritual.

10. Girls and boys receive animal gifts known as *ajif* at their naming ceremonies on the seventh day after birth.

11. By means of classical concepts of residential rules, Tuareg are most often labelled as 'patrilocal', but they could be classified as 'matrilocal' as well, since the husband lives in the tent of his wife which is seen as an extension of her mother's tent.

12. Tuareg relationship terminologies classify the mothers and sisters of *Ego*'s mother as 'mothers'. Matri- and patrilateral parallel cousins are called 'sisters', whereas cross-cousins are designated by a specific term of reference (*tabobazt*, sg. *shibobazan*), but may also be called 'sisters', both as a term of reference and a term of address.

13. The distinction of 'public' and 'domestic' is only used here by way of commodity, but it is not adequate for Tuareg society. See Claudot-Hawad and Hawad (1987) for the distinction Tuareg make between the 'domesticated interior', considered as a female domain, and the 'domesticable exterior', a necessary counterpart seen as a male domain.

14. The Inesleman women of the Azawagh are, however, an exception, since they cover their faces occasionally. I will come back to this point later in this article.

15. Tuareg have a specific concept of 'self-control' that has nothing in common with Western psychoanalytical ideas of repression. It is based, on the contrary, on the idea of transforming emotions into consciously perceived feelings in order to enable the person to adjust his/her reactions and behave adequately to the situation. A fetus acquires the ability to feel emotions as soon as it receives its first 'soul' four months after conception; consciousness (*anesfrey*, literally 'mutual feeling/suffering') emerges only at birth, when the newborn child receives its second 'soul' through its first breath. Breastmilk is linked to this divine breath and is meant to sharpen the baby's consciousness as a first step in a long process of learning 'self-control'. For this reason, an infant is breastfed by several women of various social strata (see Walentowitz 2003, 2004b).

16. This is a summary of a story published in Tuareg language by Khamidun (1976: 46–9).

17. See Figeureido-Biton (2001) for the gendered dynamics of education in Tuareg society.

18. A foreign visitor arriving in a Tuareg camp will not direct himself to the tent that looks wealthy or is surrounded by numerous livestock, but to the tent that exposes a bowl of hospitality (*tazawat*) in the tree next to it. This large wooden bowl is only possessed by people who are 'able to fill it', meaning that they usually receive many guests.

19. These dynamics of kinship might be compared with the tug of war game where each camp tries to pull the rope (cognatic kinship lines) on its own side (paternal apical ancestor).

20. Inesleman women have numerous stories of veiled encounters with men of foreign origin or of lower social status, whom they finally identify as 'not being men', thus dropping their veils and leaving their faces uncovered in their presence.
21. See also Weiner's analysis of the connections between cloth and women's reproductive roles (1989).
22. Although fattening and force-feeding provokes numerous health problems like cardiovascular diseases, old ladies often state that women were much more healthy in the times when they where fattened than nowadays.
23. Fattened women like to say that, anyhow, men do not really make the difference, whereas they themselves do!
24. Archaeologist Cathy Spieser (2006) draws a parallel between this Tuareg symbolism and that of the *imiout* in Ancient Egypt. *L'imiout*, a recipient made of jackal skin, represents the matrix and is associated with regenerative power in funeral mythology. A similar symbolism of protective skins is associated with Athena's *aegis* shield, a word that possibly derives from the Berber word for 'goat' (Camps and Chaker 1996).
25. In Tuareg cosmology, spirits (*kel essuf*) inhabit the undomesticated outside world. They are supposed to live and procreate like humans, but perform everything in a reverse way.
26. For a detailed description and analysis of this complex ritual including photographs see Walentowitz (2003).

References

Ag-Khamidun, A. 1976. *Imayyan d-elqissaten en-Kel-Denneg (Contes et récits des Kel Denneg)*. Copenhagen: Akademisk Forlag.

Barnard, A. and A. Good. 1984. *Research Practices in the Studies of Kinship*. London: Academic Press.

Bernus, E. 1998. 'Gavage (*adanay*) chez les Touaregs Iwellemden Kel Denneg', *Encyclopédie Berbère*, vol. II. Aix-en-Provence: Édisud, pp. 2996–98.

——— 1998. 'Gavage (*adanay*) chez les Touargs Iwellemden Kel Denneg', *Encyclopédie Berbère*, vol. II. Aix-en-Provence: Édisud, 2996–8.

Bernus, S., P. Bonte, L. Brock and H. Claudot (eds). 1986, *Le fils et le neveu. Jeux et enjeux de la parenté touarègue*. Paris/Cambridge: Maison des Sciences de l'Homme/Cambridge University Press.

Bonte, P. 2000. 'Les lois du genre. Approches comparatives des systèmes de parenté arabes et touaregs', in J.-L. Jamard, E. Terray and M. Xanthakou (eds), *En substances: Textes pour Françoise Héritier*. Paris: Fayard, 135–56.

Bouman, A. 2003. 'Benefits of Belonging: Dynamics of Iklan Identity, Burkina Faso', Ph.D. dissertation. University of Utrecht.

Bourdieu, P. 2000 (1972). 'La parenté comme représentation et comme volonté', in P. Bourdieu, *Ésquisse d'une théorie de la pratique*. Paris: Éditions du Seuil, 83–215.

———, J.-C. Chamboredon and J.-C. Passeron. 1968. *The Craft of Sociology : Epistemological Foundations*. New York: Aldine de Gruyter.

Camps, G. and S. Chaker. 1996. 'Égide', *Encyclopédie Berbère*, vol. XVII. Aix-en-Provence: Édisud, 2588–9.

Claudot, H. 1984. 'Femme idéale et femmes sociales chez les Touaregs de l'Ahaggar', *Production pastorale et société* 14: 93–105.

——— and Hawad, 1987, 'Le lait nourricier de la société ou la prolongation de soi', in M. Gast (ed.), *Hériter en pays musulman*. Paris: CNRS Éditions, 128–55.

Claudot-Hawad, H. 1990. 'Honneur et politique: les choix stratégiques des Touaregs pendant la colonisation française', in H. Claudot-Hawad (ed.), *Les Touaregs. Exil et résistance, REMMM* 57: 11–47.

——— 1994. 'Cosmogonie touarègue', *Encyclopedie Berbère*, vol. XIV. Aix-en-Provence: Édisud, pp. 2137–38.

——— 2000a. 'Honneur, II. Chez les Berbères du Sud (Touaregs)', *Encyclopédie Berbère*, vol. XXIII. Aix-en-Provence: Édisud, pp. 3498–501.

——— 2000b. 'Captif sauvage, esclave enfant, affranchi cousin ... la mobilité statutaire chez les Touaregs (Imajeghan)', in M. Villasante-de Beauvais (ed.), *Groupes serviles au Sahara*. Paris: CNRS Éditions, pp. 237–68.

——— 2002. 'Noces de vent: Épouser le vide ou l'art nomade de voyager', in H. Claudot-Hawad (ed.), *Voyager d'un point de vue nomade*. Paris: Paris-Méditerranée, 11–36.

Conte, É. and S. Walentowitz. 2006. '"Ties of Milk" and the Grammar of Closeness in Muslim Contexts', Paper Presented at the Sixth Biannual Conference of the European Association of Social Anthropologists, Bristol.

——— 2010. 'Kinship Matters. Tribals, Cousins, and Citizens in Southeast Asia and Beyond', *Études Rurales*, 184: 217–48.

Drouin, J. 1984. 'Sois belle et subtile ou l'art des connivences chez les Touaregs', *Littérature orale arabo-berbère* 15: 1–30.

Figueiredo-Biton, C. 2001. 'Conceptualisations des notions de chaud et de froid. Système d'éducation et relations hommes/femmes chez les Touaregs (Imedédaghen et Kel Adagh, Mali)', Ph.D. dissertation. Paris: École des Hautes Études en Sciences Sociales.

Godelier, M. 1982. *La production des grands hommes: Pouvoir et domination masculine chez les Baruya de Nouvelle-Guinée*. Paris: Fayard.

Norris, H. 1975. *The Tuaregs: Their Islamic Legacy and Its Diffusion in the Sahel*. Westminster: England Aris and Phillips.

Popenoe, R. 2004. *Feeding Desire: Fatness, Beauty and Sexuality among a Saharan People*. London: Routledge.

Spieser, C. 2006. 'Vases et peaux animales matriciels dans la pensée religieuse égyptienne', *Bibliotheca Orientalis* 63(3/4): 220–33.

Tauzin, A. 1981. 'Sexualité, mariages et stratification sociale dans le Hodh mauritanien', thèse de 3ème cycle. Paris: Écoles des Hautes Études en Sciences Sociales.

———— 2001. *Figures du Féminin chez les Maures (Mauritanie): Désir nomade.* Paris: Khartala.

Walentowitz, S. 1999. 'L'Ignorance des Inesleman de la Tagaraygarayt par le pouvoir colonial: L'élite politique des "religieux" mise aux marges de l'histoire', *Nomadic People* 2(1/2): 39–64.

———— 2002. 'Partir sans quitter: rites et gestes autour des déplacements féminins chez les Inesleman de l'Azawagh', in H. Claudot-Hawad (ed.), *Voyager d'un point de vue nomade.* Paris: Éditions Paris-Méditerranée, pp. 37–52.

———— 2003. 'Enfants de l'Autre, enfants de Soi. La construction symbolique et sociale des identités à travers une anthropologie de la naissance chez les Touaregs (Kel Eghlal et Ayttawari Seslem de l'Azawagh, Niger)', Ph.D. dissertation. Paris: École des Hautes Études en Sciences Sociales.

———— 2004a., 'L'enfant qui n'a pas atteint son lieu. Soins aux prématurés chez les Touaregs de l'Azawagh', *Revue l'Autre* 5 (2): 227–41.

———— 2004b, '*Ego* et *Alter* ou comment la parenté fait corps avec la personne chez les Touaregs de l'Azawagh', in F. Héritier and M. Xanthakou (eds), *Corps et affects.* Paris: Odile Jacob, 169–85.

———— 2006. 'Tout un monde au creux d'un plat. Propos sur la grande écuelle des Touaregs', *Journal des Africanistes* 76(1): 105–22.

———— in press. 'Germanité, alliance, filiation. Dynamiques autopoïétiques de la parenté touarègue, in P. Bonte, E. Porqueres i Gené and J. Wilgaux (eds), *L'argument de la filiation aux fondements des sociétés européennes et méditerranéennes.* Paris, Éditions de la Maison des Sciences de l'Homme, pp. 107–37.

Weiner, A. 1989. 'Why Cloth? Wealth, Gender and Power in Oceania', in A. Weiner and J. Schneider (eds), *Cloth and Human Experience.* Washington, DC: Smithonian Institution Press, 33–72.

———— 1992. *Inalienable Possessions: The Paradox of Keeping-while-giving.* Berkeley: University of California Press.

———— 1995. 'Reassessing Reproduction in Social Theory', in F.D. Ginsburg and R. Rapp (eds), *Conceiving the New World Order: The Global Politics of Reproduction.* Berkeley, Los Angeles/London: University of California Press, 407–24.

Worley, B. 1991. 'Women's War Drum, Women's Wealth: The Social Construction of Female Autonomy and Social Prestige among Kel Fadey Twareg Pastoralists', Ph.D. dissertation. Columbia University.

———— 1992, 'Where All the Women Are Strong: Wrestling Caps a Desert Tribe's Infant Ceremony', *Natural History* 11: 54–65.

Chapter 5

CHILDBEARING, BREASTFEEDING AND BODY WEIGHT IN TANZANIA:
THREE BODIES, THREE INDIVIDUALS, MANY DIFFERENT INTERRELATIONS AMONG THE WAGOGO (CENTRAL TANZANIA)

Mara Mabilia

Introduction

My field research among Gogo mothers in the village of Chigongwe in the Dodoma District (Central Tanzania) was promoted to understand the precociousness of malnutrition in children during the period devoted to breastfeeding.[1] The problem of malnutrition among children under five was very serious in the whole district. During the clinic for the monthly weighing in the dispensary of the Chigongwe village, I received confirmation of the percentage present in the area: 9 per cent of the children were severely undernourished (the phenomenon is clinically known as *marasmus* and *kwashiorkor*[2]). My basic objective was to detect any traditional cultural breastfeeding and weaning practices that could be detrimental to children's growth and health. The situation in the field, presented in these terms, was very far from suggesting that fatness was an issue. Yet, paradoxically, the fact itself that the reality under study was so visibly alien to the subject at issue – fatness – was what prompted me to further reflect on aspects closely related to body fatness, whose significance and representation in the collective imagination are familiar even to a community so plagued by food shortage. The rather poor adult daily diet (the staple meal being made up of stiff sorghum or millet porridge and *ilende*, a green-leaf vegetable available after the early rains), the frequent droughts following the periodic failure of a single rainy season characterized by being extremely erratic for about five months a year: all of this was clear evidence of widespread malnutrition rather than over-nourishment among the population. As for children, given the dangers to their survival during the early years, a dramatic constant in many sub-Saharan areas, what must be considered is the feeding practices and the beliefs associated with their growth, their needs, their being regarded as particularly vulnerable and at the mercy of adult behaviour.[3] That is why a newborn's body, its state and its tangible reactions become the object of careful observation aimed at reading in it not only its good or bad health state but also the reflection of its parents' behaviour, whether or not respectful of traditional rules.

If we think of 'fatness' not only as the result of a diet, of individual greed or organic dysfunction, but also as the definition of a shared sense of what is beautiful, ugly, healthy, or sick, then the dimension of the body, the vessel and vehicle of manifold meanings,[4] needs to be considered through a process of contextualization which translates the same categories into different interpretations and meanings. In other words, a situation such as the Wagogo's, where being fat is a very remote possibility, does not in itself entail the absence of a representation of the body characterized by the binary opposition thin versus fat, and others related to it (think of healthy versus sick, beautiful versus ugly), an opposition which, once processed, can take on different meanings in different contexts. For example, there is, as we shall see, 'beautiful' fat denoting good health, as opposed to 'ugly' fat denoting bad health in the child.

Analogously, an adult's thinness due to daily absence or scarcity of food is experienced as different from what is recognized as 'usual' thinness during a time of good harvest. In the former instance, although the difference is not particularly marked in body shape, thinness takes on a negative connotation attaching to the individual, so that, for example, he/she refuses to be photographed. The women's frequent request not to be photographed during a time of food scarcity because they were 'too thin' highlighted a sensitivity to the discrimination between comparative thinness and fatness that could escape a European's notice.

As early as the first field researches, anthropologists became aware of the importance of food in community life. Radcliffe-Brown's study on the Andamans (1922) and Malinowski's (1922) on the Trobriands, although from different angles, caught the importance of food beyond its survival value. Only with Audrey Richards (1932), however, was food suggested as an indicator of social relationships, opening up an interesting and fruitful study field for anthropologists. Food and nutrition make the body the receptacle of manifold meanings transcending its physiological and biological processes.

The interesting aspect of the questions that involve the body is that in all societies, and in every epoch, it is the centre of a complex network in which a variety of representations, beliefs and systems of symbols are intertwined, making the body a field of comparison and dispute between subjectivity and social rules, between its physical dimension and the ideal conceptualization of its shape, between its correct or incorrect use in behaviour. Therefore the body's morphology, physiology, organic specificity and also its reproductive capacity and adaptation to the environment, are "natural resources" for the mental activity of human beings.[5] 'The human body – as Victor Turner wrote – is constantly being altered by natural and cultural processes.'[6]

After these preliminary remarks, let me return to my field to consider the human bodies involved in the reproductive processes – conception, pregnancy, childbirth – as related to the practices involved in feeding the fetus and newborn. In that context, the well-being of the baby's body in all the different phases of birthing and parenting becomes the locus upon which the social and moral values of a mother, but also of a father, are portrayed through its weight, size and shape.

The different outlines of nutritional processes

During pregnancy

In the Gogo mind, during the physiological occurrences of childbearing and breastfeeding the woman's body is not an independent entity. It is closely connected to the husband's body and the fetus' body during pregnancy, and to the newborn's body, and again, in a sense that we shall see, to the husband's body during lactation.

How is a woman's body related to her baby's and husband's bodies? First, in a physiological sense.

During childbearing the centrality of a woman's body is due to its being a bearer of 'another weight in the stomach'. Nourishing the fetus for the whole pregnancy period is the main task of a future mother. But in this task she is not alone. Her husband is also involved.

While the Wagogo think that one act of sexual intercourse is sufficient to establish pregnancy, it is also a common belief that sexual activity is useful for the development of the fetus. Both parents can contribute to the process of feeding the fetus through repeated intercourse. The father's sperm feeds the fetus through sexual intercourse at the same time as the woman's blood feeds the fetus day after day: which by the way is how her lack of menstrual periods is interpreted. It is interesting to note that women named sperm indifferently through the terms 'water' or 'blood' when they referred to fetus-feeding sexual intercourse during pregnancy. Much more rarely did the term 'milk', in the same circumstances, replace either. These different namings may sound disconcerting if one thinks of the diversity of the fluids involved in referring to sperm: water, blood and milk. In actual fact, one is not confronted with conceptual confusion on the part of the women, but with clear terminological synonymy aimed at connoting a 'vital principle'. All the terms alternatively used to mean the same fluid – sperm – proved to be synonymous in the context, all of them denoting fluids fundamental for both reproduction and life itself.

The couple's cooperative feeding of the fetus, which is in turn the combination of the male seed that met the female 'seed' in the woman's womb, highlights their continuous pooling – 'mixing' – action, aimed at the creation and development of the fetus.[7] The event of impregnation and the repeated sexual (i.e., nutritional) acts, the combination in the fetus first of seeds and then of different fluids, will – as can be seen below – foster important after-birth balances which the newborn's body will be dependent on for a long time, and which will prove fundamental for its survival and growth.

The quality of this repeated intercourse, where the responsibility of a good father towards his future child has been outlined, ceases after the first three to four months of pregnancy. The Wagogo think that this is the period necessary for the development of a fetus' body structure. A minute but complete body has developed. From now on, the baby needs only the maternal fluids, blood and colostrum, to grow, and sexual intercourse is forbidden.[8]

In their traditional concept of reproduction, the Wagogo may be seen as making the same distinction as Darwin suggested between development and growth in biology. George Canguilhelm interprets this distinction as opposing the embryo to the adult in the twin aspects of structure and size. In this sense human beings go on 'growing' after they have ceased to 'develop'.[9] Thus, during early gestation the father's cooperation is required for the fetus' development, after which time his task is over, as development is concluded. From that time on the fetus only needs to be fed to grow, and for this the mother is solely responsible.[10]

But what happens if the father is absent during these early months of his wife's pregnancy? What will his baby be like? At birth it will be thin and, probably, smaller. At any rate, it cannot be 'fat' (in the simple sense of 'not thin'). However, since a baby is born at all, there would seem to be contradiction with the belief in the necessity of the father's nutritional contribution. The women parry the contradiction by highlighting the 'not-fatness'. In this sense, having a 'fat' newborn baby confirms both the solidity of traditional beliefs and the father's correct behaviour. Let children be born even though, during pregnancy, mothers have no intercourse with husbands owing to the latter's absence: the parry is an attempt to account for a belief which, at the time of my presence in the field, must have coexisted with changes taking place especially in the economic organization of the community.

During my research the men not only left the village to look after cattle, but altogether migrated to towns or other villages in search of jobs.[11] Their absence from the village could last for days, weeks or even months, and they could be so far from their villages as not to be able to 'feed the fetus' regularly or come back promptly if necessary. The older women, for example, complained of this absence especially during the early months of the babies' lives, when the presence of their fathers was necessary to make decisions and take action for the newborns' well-being. This is one of the very few cases in which I witnessed women's criticism of men's behaviour.

Be that as it may, however indecisive the father's behaviour has proved to be, strict emphasis is still put on the absolute prohibition for the woman to have intercourse after the period devoted to the fetus' development. If she does not observe this rule, the newborn's body could be 'grey', which in practice means covered in an oily substance. While 'being born fat' is causally linked with the fetus' development during the first months of gestation, which both parents are involved in, the newborn's grey body is the sign that the mother has had sex only for her pleasure, while her whole body should have been devoted to her baby's growth. The newborn's grey body is the sign that the traditional rule has been broken: a sign of a future mother's shameful behaviour.

When such a birth occurs, the women who have helped the mother to deliver refuse to wash the newborn. Her mother-in-law must wash it. This is a gesture of tacit but stern rebuke, a source of great embarrassment for a mother.[12] Although the unlawful behaviour was obviously cooperative, these blaming and shaming attitudes are addressed exclusively to women, while no charge whatever is levelled at men. And it is paradoxically the women who make themselves the expression of a rule where the man's role is invisible.

These beliefs and their close connections are aspects of the cognitive experiences which, as they are reinforced day by day through communal life, deeply affect the behaviour and intimate feelings of both women and men. Here, we see how the physical link between the fluids of two bodies and their mixing in the nutrition of a third body becomes a cultural construct which defines the society's ideas of womanhood, manhood and different gender roles. The same construct acts as a benchmark by which the quality of the prospective parents' behaviour is judged, on the evidence of the baby's appearance at birth in terms of 'fat' versus 'thin' and 'clean' versus 'dirty'.

During breastfeeding

After delivery the baby's body becomes the attention centre not only of its mother, of course, but also of other women in the family and in the community: an object of careful and continuous observation, exactly the opposite of what its mother's body was during pregnancy. In fact, when a woman is pregnant, the people around her must pay no attention to her state and, as for herself, the future mother must protect her pregnancy from bad and envious looks, she must not even talk about it. Only the women close to a future mother know about her pregnancy. The correct behaviour is to ignore her physical condition, the external change in her body shape. So a pregnant woman continues with her working patterns and usual diet: no special food, no exemption from domestic and agricultural work.[13]

As I said, in a rural village of central Tanzania, placed in an environment not exactly favouring the survival of human beings and animals, the crucial and critical point is not, as in our society, a surplus of food that is a cause of fatness, but the problem of under-nourishment in adults and children. The problem, that is, lies not in food but in the constant lack of it. When the Wagogo are lucky, they eat – every day, once a day – the same thick porridge and vegetables.

But even in this difficult environment, food is a 'tool for thinking'.

What does being fat or thin mean for an adult in this environment? As specified in the introduction, despite the cyclical lack of food, the idea of a fat body is not absent in adults. But let us not think of our own conception of 'fat', or even 'plump': a Gogo 'fat' body is closely connected to the idea of having food beyond survival, i.e. available for exchanging or offering. The awareness of not having food for oneself and others is shameful.

So not having food is a depressing condition for both the body and spirit. Being thin, being perceived as a fatless body because one has no food is a doubly depriving condition. First, it threatens the survival of the individual's physical, i.e. 'object' body. Second, it threatens the survival of a 'subject' body, i.e. of the body as a relational subject. Offering, sharing and exchanging food are not only nutritional actions. They are wholly associated with the capability of creating and/or maintaining social ties, on the basis of which a well-fed ('fat') body is a signal to be manifested, whereas a thin one is not. One need not be ashamed if one is thin on account of illness,[14] but one must be ashamed if one is thin for lack of food. Here, being thin is as a symptom, above all, of a lack of interrelationships which involve mutual give-and-take. Being part of a network of exchanges is culturally vital, as testified by numerous anthropological and sociological studies of reciprocity.[15] So being 'fat' or 'not fat' is a symbol whose cultural, social and psychological meanings go well beyond the body's physical shape.

Despite the above-mentioned scarce dietary resources, some foods are forbidden during pregnancy, and more generally to every fertile woman. The most important taboos are cow's udder and the bits of meat left in the hide of skinned cows or goats. The former is believed to prevent the woman from having milk in her breasts and the latter to cause a bleeding delivery. These taboos are the cultural heritage of not-far-off days when livestock in every household made the struggle for life less precarious. Eggs or honey are also prohibited as they could cause the child to be born hairless.

While the correct behaviour towards a pregnant woman is to ignore her physical condition, the external change in her body shape, the attitude to the newborn is totally different, its body being made the object of constant attention, as it is the symbol of the baby's health status, and of much more, as we shall see. The baby's vitality is another important signal. Not only is a mother is proud of the weight that she feels when her baby is tied to her back, delighted by the skin folds on the thighs that she kisses, but the baby's liveliness, its ability 'to look round itself' is also the object of attention and praise, as the women told me. These different attributes can be summed up by the phrase 'fat baby' in which 'fat' is synonymous with a baby in good health, a baby that is growing well.

This general state of well-being and growth that a mother reads in the signals sent by the baby's body, its shape and mobility, can change, sometimes suddenly, through high fever or acute diarrhoea, at other times more slowly but treacherously, when modest but persistent diarrhoea can lead to a disastrous outcome.

A baby's body may have a florid appearance but, when weighed or simply lifted, be felt to be 'weightless', i.e. not corresponding to what one would expect. The child's body sends a falsely healthy message when it is, in fact, weightless: its fat body is 'falsely fat' and therefore a 'bad fat body': fat may thus be a misleading 'marker'. That happens when the baby suffers from a serious form of malnutrition, called *kwashiorkor* by Western medicine. A swollen body can be taken for a fat body, at first sight.

What are thought to be the causes of this 'false fatness' or, more generally, of the loss of fat in a reality in which the idea of a baby's fat body is synonymous with a baby in good health, a baby that is growing well?

A mother's first and foremost task is to feed her newborn baby with 'good' food to make it grow up in 'good' health. The only and best food for a newborn baby is her breast milk. Through this fluid a mother continues to be the source of her baby's nourishment, a task that started with pregnancy. Breast milk replaces her blood in feeding her baby's body, instilling force, strength, and gives it fat. This fluid has many special qualities for the baby's needs, but in the mind of the Gogo mother it can also be a reactive substance that can change from 'good breastmilk' to "bad breastmilk'.[16]

When and why does breast milk change to become 'not good' for a lactating baby? The causes may be various, but by far the most important one occurs when a mother is not following all the rules that can help her maintain her offspring in good health. The paramount rule for this purpose is post-partum sexual abstinence: a taboo that she must follow for the whole period of breastfeeding. Therefore, a mother must observe sexual abstinence refusing to accomplish her duties as a wife for the whole breastfeeding period. Being a mother, a wet-nurse-mother, is her all-important role as long as she gives her breast milk to her child.

Her husband, the baby's father, cannot have sexual intercourse with any woman at all for three to four months after birth, then he must continue to abstain from giving his sperm to the lactating mother. The Wagogo think that this avoidance of sexual intercourse between the newborn's parents is required to safeguard the physiological equilibrium which binds the three bodies: the woman's, the man's and their child's. This balance is the result of the combination of the two different kinds of seeds (female and male seeds) from which 'the baby jumps out'. It continues during pregnancy through the male and female fluids for the development and growth of the fetus' body, and is the same equilibrium on which the healthy and strong growth of the newborn's body can depend, and towards which a mother and a father have different duties. During the whole period of breastfeeding a mother gives her fluid (breast milk) to feed her baby and the father must not give his fluid (sperm) to his wife for the same purpose: giving health – growth, fat, vitality – to their baby.

Thus women ascribe the onset of some severe forms of diarrhoea striking a milking newborn to the violation of this rule. Sexual activity, in their view, produces in the mother's body a new mix of male and female fluids which now alters the balance created at the newborn's conception and on which it still depends for its well-being.

The baby's body, actually the weakest and most vulnerable subject in the above mutual interdependence, sends a signal that the primary equilibrium has changed, the signal consisting of a persistent or acute diarrhoea which harms its body and, ultimately, puts its survival at risk.[17] Sudden and worrying loss of weight or 'false' fat will cause its death in the absence of timely intervention in accordance with traditional rules. Diarrhoea, a destructive fluid, will drain away the fat and, with it, the baby's health.

In these circumstances, the mother's breast and milk are thought of as almost autonomous entities, 'bad' milk scarcely affecting the mother's health and not being produced by her own physical body state. Furthermore, feeding the mother well does not make her milk any better.[18] In short, 'milk' is a 'separate system'. As a 'separate system' breast milk is a moral barometer and diarrhoea, with its serious effects on the baby's body, fat loss first of all, is the result of infringing social and moral codes. Then, the mother's breastmilk, this special fluid that comes from her body, can be – depending on circumstances – a 'producer' or a 'destroyer' of fat.

Here the infringement of sexual taboos makes the link between these particular fluids, breast milk and diarrhoea, extremely susceptible to external behaviour. The baby's diarrhoea, since a baby is not a social human in its own right, must be seen as an extension of the mother's fluids, in this case her milk, which is the only link between the two beings.

A newborn's grey body, a fatless newborn, a baby's weight loss, a weightless fat baby, a baby beginning to look like a wrinkled-skinned old man, are all different appearances of baby's body, witnesses to concealed social disorder. The woman has violated the social order by allowing herself to be penetrated, by taking in her husband's sperm. She has broken rules which protected the balance that had been established among the three bodies through peculiar sexual acts having as their purpose the newborn's well-being. Unlawful mixes have produced results detrimental to her child's health. And this is of course a clear example of gender inequality: it is the woman's responsibility to preserve her role as a wet nurse mother by denying herself to her husband as a wife. No provision is made for her role as simply a woman!

The example outlined here of the Gogo mothers confirms two aspects closely linked to the body. The first aspect highlights its peculiar versatility as an object of symbolization and conceptualization regarding its biology and physiology: a continuous exchange of meanings between its nature and its being the object of interpretations, as noted by Mauss (1936) when he wrote that the study of body techniques cannot be separated from the study of symbolic systems.[19]

The second aspect confirms the body as the instrument through which human societies exert power and control.[20] Here, it is the woman's body that is kept under close scrutiny through the observation of another body, that of her baby. A body to be kept healthy, i.e. 'fat', a fatness that melts when the mother's body, to which it is linked for its survival through lactation, breaks the rule of sexual abstinence. What melts away with her baby's fat is her reputation as a 'good' mother, and the recognized hierarchy of the genders. And, ultimately, what gets melted is the credibility itself of social rules.

Conclusion

The example of Gogo mothers, alongside the broader picture of Gogo society that goes with it, points to a meaning of 'fatness' – as referred to a breastfed baby's body – which strongly relates to the cultural environment. This would sound like nothing new, were it not for the intriguing ways in which the baby's body acts as a 'significant' for the whole community in the breastfeeding context: that very circumstance of the mother-baby relationship which one would tend to consider as the epitome of intimacy and exhaustiveness.

An adequate understanding of the breastfeeding process requires consideration of the close interrelationships among its three main actors (mother, father and fetus/baby) within a system of norms, prohibitions and beliefs – including, for the woman, total *post-partum* sexual abstinence – aimed at preventing alteration in the milk's quality and consequent damage to the baby's health. Inside this cultural framework, a 'fat' baby – which is synonymous with a healthy one – acts as the indicator of correct behaviour by the father and, especially by the mother-wife in her capacity as a wet-nurse. The gender-linked unbalance is obvious.

Judging a baby as 'fat' or 'thin' implies seeing its body as the projection of its parents' respect or neglect of norms: as the source of approval or stigma. The concepts of a 'fat' or 'thin' baby, in the very context of a community where coping with the problem of procuring food for both adults and children is a major concern, point to their being used as ordinative signs of men's and women's behaviour whose control by society is thought of as a basic condition of the reproductive success of both each single kinship group and the community as a whole.

Notes

1. On my research, see references.
2. *Mother and Child Health Report* (MCH) 1989/90.
3. The link between disease and bad behaviour of one or both parents has been widely witnessed in anthropological literature.
4. There is a rich literature on the topic. We will limit ourselves to a classic paper, Scheper-Hughes and Lock (1987).
5. Douglas 1979.
6. Turner 1987.
7. For cultural details of fecundation see Mabilia 1996a, 2000, 2005.
8. As mentioned in a previous publication, I often received contradictory answers about the issue of the presence of colostrum at the beginning of pregnancy as a nutritional substance for the baby that is 'in the mother's stomach'. There is, anyway, a fairly widespread belief that was repeated to me: 'he sucks inside (the (mother's) stomach, until he comes out to light' (Mabilia 2005: 70). It is possible that colostrum, as

well as the mother's blood, is conceptually assumed as synonymous with 'vital fluid', as it helps the baby expel the meconium.

9. Canguilhem 1970: 115.

10. Medical science sets the end of organogenesis at the twelfth week of gestation. During the remaining time the fetus grows and completes its functional structure.

11. If the number of livestock has been increasing for several years, they was not uniformly distributed throughout the households. Many families have not bovines and their survival depend upon the agriculture activity which is subject to the caprice of a single and often elusive rainy season (Mabilia 2005: 9).

12. Mabilia 2005.

13. Mabilia 2005: 70.

14. At the time of my research, 1989 to 1992, HIV-AIDS was making its appearance in the city of Dodoma on account of male mobility. I had therefore no way of considering this epidemic during my field work. However, I do remember that in the case of some illnesses the subject's responsibility is recognized by the community as a stigma.

15. Reciprocity and gift-giving have been the object of a long anthropological and sociological tradition, featuring authors such as Mauss (1924–25), Malinowski (1922), Lévi-Strauss (1947, 1965), and, more recently, Douglas (1990), Godbout (1992), Caillé (1998), and Godelier (1999).

16. Mabilia 2000, 2001.

17. Brandt and Rozin (1997) showed that the idea of a close connection between breach of rules and well-being of individuals is widespread well beyond African societies.

18. As her body, size and weight, during the length of lactation, are not objects of attention and consideration, neither are there any special diets or foods. Only after delivery is soft porridge eaten by the puerpera and it is a sign of warm welcome to offer a morsel of it to a crying baby.

19. Mauss 1936 (Italian trans., 1965b: 392).

20. Foucault 1979.

References

Brandt, A.M. and P. Rozin (eds). 1997. *Morality and Health: Interdisciplinary Perspectives*. New York: Routledge.

Caillé, A. 1998. *Le tiers paradigme. Anthropologie philosophique du don*. Paris: Éditions la Découverte.

Canguilhem, G. 1970. *Études d'histoire et de philosophie des sciences*. Paris: Vrin.

Corbin, A. 25 January 2005. 'La vera storia del tuo corpo', in *La Repubblica*, my trans., p. 37.

Douglas, M. 1979. *I simboli naturali*. Torino: Einaudi. (Original title, *Natural Symbols*. London: Penguin Books Ltd, 1973.)

———— 1990. 'Foreword', in M. Mauss (ed.), *The Gift: The Form and Reason for Exchange in Archaic Societies*. London: Routledge, pp. vii–xviii.

Foucault, M. 1979. *Discipline and Punish: The Birth of the Prison*. New York: Vintage. (Italian trans., *Sorvegliare e punire. Nascita della prigione*. Torino: Einaudi, 2005).

Godbout, J.T. 1992. *L'esprit du don*. Paris: Éditions la Découverte.

Godelier, M. 1999. *The Enigma of the Gift*. Cambridge: Polity Press.

Lévi-Strauss, C. 1947. *Les structures élémentaires de la parenté*. Paris: Presses Universitaires de France (Italian trans. *Le strutture elementari della parentela*. Milan: Feltrinelli, 1969).

———— 1965. 'The Principle of Reciprocity', in L.A. Coser and B. Rosenberg (eds), *Sociological Theory*. New York: MacMillan, pp. 77–86.

Mabilia, M. 1996a. 'Beliefs and Practices in the Breastfeeding and Weaning among the Wagogo of Chigongwe, Dodoma Rural District, Tanzania: I. Breastfeeding', *Ecology of Food and Nutrition* 35(3): 195–207.

———— 1996b. 'Beliefs and Practices in the Breastfeeding and Weaning among the Wagogo of Chigongwe, Dodoma Rural District, Tanzania: II. Weaning', *Ecology of Food and Nutrition* 35(3): 209–17.

———— 2000. 'The Cultural Context of Childhood Diarrhoea among Gogo Infants', *Anthropology and Medicine* 7(2): 191–208.

———— 2001. 'Allattamento come dono? Un caso etnografico', *DiPAV Quaderni, Semestrale di Psicologia e Antropologia culturale*, no.2, Franco Angeli Editore, pp.115–36.

———— 2005. *Breast Feeding and Sexuality*. Oxford: Berghahn Books.

Malinowski, B. 1922. *Argonauts of Western Pacific*. New York: Dutton.

Mauss, M. 1924 and 1925. 'Essai sur le don', *Année Sociologique*, II. Id. Paris: Sociologie et Anthropologie, Puf, 1950, pp.. 143–279 (Italian trans., 'Saggio sul dono', in Id., *Teoria generale della magia e altri saggi*. Torino: Einaudi, 1965a, pp. 153–292).

———— 1936. '*Les techniques du corps*', *Journal de Psychologie* 32(3/4). (Italian trans., 'Le tecniche del corpo', in *Teoria generale della magia e altri saggi*. Torino: Einaudi, 1965b, pp. 384–409).

MCH. 1989/90. *Mother and Child Health Report*, unpublished Annual Record. Dodoma Hospital.

Radcliffe-Brown, A.R. 1922. *The Andaman Islanders*. Cambridge: The University Press.

Richards, A.I. 1932. *Hunger and Work in a Savage Tribe*. New York: Meridian Books.

Scheper-Hughes, N. and M.M. Lock. 1987. 'The Mindful Body: A Prolegomenon to Future Work in Medical Anthropology', *Medical Anthropology Quarterly* 1(1): 6–41.

Turner, V. 1987. 'Bodily Marks', in M. Eliade (ed.), *The Encyclopaedia of Religion*, vol. II. New York: MacMillan.

Chapter 6

THE 'OBESITY CYCLE':

THE IMPACT OF MATERNAL OBESITY ON THE EXOGENOUS AND ENDOGENOUS CAUSES OF OBESITY IN OFFSPRING IN THE UNITED KINGDOM

Nicola Heslehurst

Introduction

Obesity is a major public health issue on an international level, and in the United Kingdom there are numerous ways in which governments attempt to prevent obesity from proliferating. The majority of public health initiatives addressing the prevention of obesity are aimed at preventing childhood obesity through lifestyle interventions due to the relationship between childhood obesity and the subsequent development of adult obesity (Parsons et al. 1999).

It is well recognised that children who are obese are likely to have obese parents (Parsons et al. 1999; Patrick and Nicklas 2005). Although there is no consensus on the causal relationship between parental and childhood obesity the position adopted in this paper is that there is a combined endogenous and exogenous relationship which cannot be viewed in isolation of each other. This paper will discuss the cyclical nature of obesity, the impact maternal obesity has on this cycle, discuss the relationship between obesity and fetal development, and question whether initiating obesity interventions in childhood is early enough to prevent the increasing prevalence of obesity.

The increasing prevalence of obesity is a major health issue in the U.K. having grown by almost 400 per cent in the last twenty-five years, obesity will soon surpass smoking as the greatest cause of premature loss of life (Health Committee 2004). Trends in prevalence of obesity vary, and when looking at relative rates of obesity within affluent countries increased prevalence is associated with low socioeconomic status, making it a health inequality issue, the prevention of which was high on the previous government's agenda (Health Committee 2004; Department of Health 2002). This is reflected in the Department of Health's (DH) white paper *Choosing Health* (Department of Health 2004a), which encompasses various issues relating to obesity, including the impact of obesity-related morbidities, inequalities in health, and child health and nutrition, and the U.K. Government's Foresight Programme aimed to identify a sustainable response to obesity over forty years (Foresight 2007).

Government recognition of the importance of interventions to prevent obesity at a young age has been seen with increasing school targets related to nutrition and exercise (Department of Health 2004a; Department of Education and Employment 1999; Department of Health 2002b). These include schools increasing the availability of fresh fruit, healthier breakfast club and school meals, encouraging walking to school, increased levels of exercise during and beyond school hours, and comprehensive health education within the National Healthy Schools Programme.

The importance of a healthy diet during pregnancy and infant years was also recognised, and the Healthy Start Scheme (Department of Health 2002b) aimed to tackle inequalities in health from an early age, by supplying vouchers for fresh fruit and vegetables, milk and infant formula to eligible mothers during pregnancy and breastfeeding, and to young children in low-income families. While the previous government recognised the importance of a healthy diet during pregnancy as outlined in the Healthy Start Scheme, the potential adverse effects of obese women becoming pregnant, on both the mother and the infant, was not addressed in the *Choosing Health* report, National Institute for Clinical Excellence (NICE) guidance (prior to 2010) or the National Service Framework (NSF) for children, young people and maternity services (Department of Health 2004c).

Incidence of maternal obesity in the U.K.

The 2004 triennial report into maternal mortality by the Confidential Enquiry into Maternal and Child Health (CEMACH) shows a relationship with maternal obesity and mortality, with 35 per cent of all maternal deaths from 2000 to 2002 occurring in obese women with a Body Mass Index (BMI) > 30 kg/m^2; 50 per cent higher than in the general population (Confidential Enquiry into Maternal and Child Health 2004). By the years 2003 to 2005, more than half of all maternal deaths were in overweight or obese women (BMI > 25 kg/m^2), with over 15 per cent being morbidly obese (BMI > 40 kg/m^2) or super morbidly obese (BMI > 50 kg/m^2) (Lewis 2007). Previous CEMACH reports highlight an absence of reliable height and weight data recorded, therefore it was not possible to further identify the trends in the incidence of maternal and perinatal mortality relating to obesity (Confidential Enquiry into Maternal Deaths 2001).

There is also an absence of reliable data relating to rates and trends of maternal obesity on a national or international level. However, the prevalence of obesity in women in England has risen from 16.4 per cent to 24.8 per cent between 2003 and 2005, with the highest prevalence amongst Black African (38 per cent), Black Caribbean (32 per cent) and Pakistani ethnic groups (28 per cent) (Department of Health 2005), and the prevalence of obesity in women of childbearing age is also increasing (Department of Health 2004b). Despite the absence of nationally representative statistics in the U.K. relating to obesity in pregnancy, two studies have shown similar results in the trends of obesity in pregnancy in the regions of Middlesbrough and Glasgow, with incidence of maternal obesity increasing from approximately 10 per cent to between 16 to 19 per cent over a fifteen-years period (Heslehurst et al. 2007b; Kanagalingam et al. 2005). Guelinckx et al. (2008) summarise the findings of published studies that indicate the rates of obesity in the pregnancy population on an international level. Although the findings of international studies are difficult to compare directly due to the variation in the categories used to define obesity, the differences in time periods of the published studies, and the fact that the majority of the studies included the U.S.A. and Australia, therefore not giving a true international representation, the authors conclude that the trends of maternal obesity on an international level are between 1.8 per cent and 25.3 per cent using the World Health Organization definition of obesity of a BMI > 30 kg/m^2.

The data provided in the Middlesbrough study also shows that the incidence of maternal obesity is increasing over time at a similar rate to the prevalence of obesity in women of childbearing age in the general population in England (Heslehurst et al. 2007b), leading to concerns about the effect of this on the relationship with maternal mortality as highlighted in the CEMACH reports, and other potential adverse effects of maternal obesity.

Obesity and reproductive health

Obesity has an impact on women's reproductive health, and there are health risks to both mother and her infant.

There is a relationship with polycystic ovarian syndrome (PCOS), infertility, and the success of infertility treatment (Wang, Davis and Norman 2002), whereas weight loss has been shown to alleviate these conditions and improve the success of infertility treatment (Clark 1995). There is an increased risk of mothers developing gestational diabetes (Andreasen, Andersen and Schantz 2004) and subsequent development of diabetes mellitus, an increased risk of hypertensive disorders and pre-eclampsia, and thromboembolic complications (Castro and Avina 2002; Linne 2004). There is some evidence of an increased risk of late fetal loss (Lashen, Fear and Sturdee 2004) and stillbirth (Cnattingius and Lambe 2002). CEMACH supports this finding in their perinatal mortality report, where mothers were obese in 22.9 per cent of all late fetal loss, 30.4 per cent of stillbirths, and 30.6 per cent of neonatal deaths in 2005 (Confidential Enquiry into Maternal and Child Health 2007).

Congenital anomalies have also been linked with maternal obesity. Waller et al. (2007) found that mothers of offspring with spina bifida, heart defects, anorectal atresia, hypospadias, limb reduction defects, diaphragmatic hernia, and omphalocele were significantly more likely to be obese than mothers of controls, with odds ratios ranging between 1.33 and 2.10. There is also a risk of complications throughout the delivery with the baby being macrosomic and the need for more frequent induced and operative deliveries (Andreasen, Andersen and Schantz 2004; Heslehurst et al. 2008).

Obesity has also been shown to have an impact on psychological health as well as biomedical health. The biomedical risk of PCOS and infertility associated with being overweight and obese appears to be on the increase, with the incidence of overweight and obese women attending for *in vitro* fertilisation (IVF) having increased over the years (Norman and Clark 1998). Although infertility is a biomedical effect of obesity, it also has a significant psychological impact on women who are unable to conceive, and with a number of hospitals in England refusing NHS funded IVF treatment for women with a BMI > 35 kg/m^2 is likely to add to the distress of infertility in the obese woman. The association between obesity and deprivation could also indicate an increase in health inequalities relating to fertility as a direct result of this BMI 'cut off', and add to the psychological distress to women who are unable to afford private healthcare for fertility treatment, and are subsequently unable to conceive due to their BMI status.

Social stigma is also associated with being obese (Reilly and McDowell 2003), and this could be heightened with increased interest by the media. The *Choosing Health* white paper identified that media coverage of obesity had increased dramatically in recent years, potentially validating the social acceptability of obesity-related stigma. The psychological impact of obesity in pregnancy is relatively unexplored; however there are issues with patient dignity, embarrassment, and feelings of victimisation when health care practitioners raise the issue with mothers (Heslehurst et al. 2007a).

There is an apparent lack of awareness of the impact of being obese in pregnancy among women, and health care practitioners have stated that women often have no perception of being obese themselves, potentially due to the normalisation of overweight and obesity among their peers. There is also the combined service implications and psychological implications relating to being able to physically detect the fetus on the ultrasound scan due to the obstructing fat mass, which has an impact on service as the scan cannot detect what it is supposed to, but also has consequences for the parents who cannot see the picture of their baby (Heslehurst et al. 2007a).

Health care practitioners have also expressed the difficulty in conveying the message of obesity to women when they are pregnant as it is difficult to get a balance of information about the potential risks, the requirement of additional procedures and more intensive monitoring, and the reduced choice of the mothers for their care plans such as midwifery-led care and mode of delivery, without causing additional upset and stress to the mothers when they are already considered to have high risk pregnancies. The findings of this study also highlights the frustration of health care practitioners in maternity units when it came to offering support and services towards maternal obesity as it was felt to be a public health issue. Health care practitioners felt that there was little that could be done in terms of weight reduction during pregnancy, due to potential risks to the fetus, and that any weight loss interventions need to be carried out before conception. It was felt that all that could be done during the pregnancy was to manage the care of the mother as safely as possible.

Life-cycle obesity prevention

As discussed the interventions to 'tackle' the growing prevalence of obesity in the population are predominantly aimed at the prevention of childhood obesity; a recognised potential precursor to adult obesity (Parsons 1999). However, it should be questioned as to whether addressing obesity at this stage in the life-cycle is early enough? There is an argument in life-course research that it is not just childhood and adulthood experiences that contribute towards adult health and inequalities, but that factors may predetermine this period and can be influenced by the intrauterine environment, even influenced preconception, and may also be related to the mothers own life-course experience (Barker 1998).

In the study of obesity, there are conflicting theories relating to the determinants of weight gain, where two opposing arguments are that the determinants are endogenous, or exogenous. Barker's 'Fetal Origins' hypothesis would argue that the determinants are based on endogenous factors, as elements are determined prior to conception and in the intrauterine environment during the fetal development (Barker 1990). The alternative argument is the exogenous causes of obesity that are lifestyle-related; this is determined based on the energy balance equation where energy consumption exceeds energy expenditure.

Power et al. support the theory that the maternal life-course has an impact on the health status of the offspring, as they identify short maternal stature as an independent risk factor for obesity in the offspring during childhood, and through to adulthood; short stature being related to poor life-long nutritional status of the mother (Power 2003). Mothers' birth weight has also been shown to affect fetal growth in her offspring, potentially due to multiple factors associated with nutritional supply to the fetus including mothers' body composition, life-long nutritional status, her dietary intake during pregnancy, and the transport capacity of nutrients to the placenta (Godfrey 1997).

Power et al. (2003) also showed a significant relationship between obese mothers, high birth-weight in offspring, and their subsequent development of childhood and adult obesity. This is supported by numerous other studies that show a positive correlation between maternal obesity, macrosomia, increased skin fold thickness, and development of obesity in offspring (Curhan et al. 1996a, 1996b; Larsen, Serdula and Sullivan 1990; Whitaker et al. 1998). However this appears to contradict Barker's hypothesis that poor fetal growth is due to maternal under-nutrition, and that the subsequent low birth weight of infants is the cause for adult obesity and its related disorders (Barker 1990). A systematic review of the childhood predictors of adult obesity showed that maternal obesity and weight gain during pregnancy are related to a greater BMI in childhood, and subsequent obesity in adulthood (Parsons et al. 1999). Parsons et al. also stated that this conflicted with the life-course studies carried out over the Dutch famine time period, which showed under-nutrition was positively associated with the development of obesity, supporting Barker's hypothesis.

These two conflicting theories, based on maternal under-nutrition and subsequent low birth weight leading to the development of obesity in the offspring, versus maternal over-nutrition, could be explained by Van der Meulin's theory that it is poor fetal growth in general that is the cause of the development of obesity in offspring (Van der Meulin 2002): regardless of whether this is attributable to maternal under- or over-nutrition. This theory is supported by the conclusions drawn from a systematic review of the determinants of obesity (Parsons et al. 1999).

When looking at maternal obesity as an independent risk factor for the development of childhood obesity in the offspring, the morbidities that often co-exist with obesity must also be considered. When women are obese at the start of pregnancy there is an increased risk that they may also have type I diabetes (NIDDM) or impaired glucose tolerance (IGT), and an increased risk of developing gestational diabetes mellitus (GDM) during pregnancy (Pettigrew Hamilton-Fairley 1997).

Life-course studies show that women who have diabetes during pregnancy persistently have obese offspring, and this is independent of genetic factors suggesting that the intrauterine environment is altered in a diabetic pregnancy (Whitaker et al. 1998; Breidahl 1996; Pettitt et al. 1993; Plaguemann 1997; Rodrigues 1998; Silverman 1991). The effect of weight gain during pregnancy as an independent risk factor for obesity in the offspring should also be considered. The effect of maternal weight gain is considered to be dependent on the pre-pregnancy weight (Whitaker et al. 1998), and although maternal obesity is associated with an increased risk of macrosomia, women who are morbidly obese at the start of pregnancy are also at risk of delivering very low birth weight infants due to poor maternal weight gain and increased risk of extreme pre-term delivery; both of which have a detrimental effect to fetal development causing low birth weight (Cnattingius et al. 1998), which according to Barker's hypothesis potentially leads to increased risk of adult obesity. Therefore, both excessive and inadequate fetal growth in pregnancy leading to macrosomia and low birth weight could have an impact on the obesity levels in the offspring.

The Institute of Medicine (IOM) in the U.S.A. developed gestational weight gain criteria for different BMI groups, which was based on the available evidence at the time (Institute of Medicine 1990). It stated the lower and upper limits of recommended weight gain for underweight, ideal, and overweight women in pregnancy to be 12.5–18 kg, 11.5–16 kg, and 7–11.5 kg, respectively. The recommendations for obese women, however, are more complex as the evidence was less robust and showed the highest variation for weight gain in this group. However a lower limit of 6.8 kg weight gain in pregnancy was set for this group, and the guidelines state that obese women should be encouraged to consume nutritious foods and a sufficient quantity of the essential nutrients.

These recommendations have now been updated (Institute of Medicine 2009), and the lower and upper limits of recommended weight gain for underweight, ideal, and overweight remain the same. Lower and upper limits for recommended weight gain for obese women in pregnancy have also been included at 5–9 kg. Despite the existence of the IOM guidelines, the U.K. has no established weight gain recommendations for pregnancy based on the mothers BMI, and the advice given to women during pregnancy regarding weight gain is often inconsistent, ad hoc, and dependent on the clinician in charge of care (Heslehurst 2007a).

The exact mechanisms as to why these endogenous factors appear to proliferate the development of obesity in the offspring are unknown. However there are multiple theories relating to the effects of the intrauterine environment. These theories refer to the effects of placental pathology, and maternal and fetal nutrition. One theorised mechanism is an alteration in the intrauterine environment affecting the transfer of metabolic substrate to the fetus, potentially impeding development of the structure and function of fetal organs involved in energy metabolism (Whitaker et al. 1998). Excessive weight gain during pregnancy is also thought to cause fetal hyperinsulinism, which may 'malprogramme' the fetal hypothalamus, pancreatic beta cells, and adipocytes, predisposing the infants to obesity (Dorner and Plagemann 1994).

Pederson's model theorises that the metabolic alterations *in utero* when the mother is diabetic are due to increased glucose and amino acids in the blood, having a similar affect to that which happens when the mother is obese, causing fetal pancreatic beta cell hyperplasia, hyperglycaemia and hyperinsulinaemia (Whitaker et al. 1998). The diminished insulin secretion and increased insulin resistance is theorised to lead to obesity in adulthood, and the effects of the intrauterine environment on the development of insulin receptors and stimulation of beta cells, or leptin production, may also contribute towards obesity development. The adipocyte number hypothesis refers to an increased transfer of fat fuels to the fetus when the mother is obese, or an elevation of triglycerides due to GDM, affecting the fat cell size and number potentially having long-term implications for obesity predisposition.

The fetal over-nutrition hypothesis relates to the persistence of fat rather than muscle in offspring of mothers who are obese or develop GDM during the pregnancy (Poston and Taylor 2007). This is potentially due to the high maternal glucose, free fatty acid, and amino acid plasma concentrations resulting in over-nutrition of the fetus, which may permanently change the appetite control, neuroendocrine functioning, or energy metabolism in the developing fetus, leading to obesity in later life (Lawlor and Chaturvedi 2006). Lawlor and Chaturvedi discuss how maternal obesity may be the prime factor in fetal over-nutrition due to the high plasma concentrations of glucose and free fatty acids, and the relationship with glucose intolerance and insulin resistance.

Further endogenous relationships between the fetal environment and the impact on the development of obesity-related behaviour may be linked to dietary preferences established *in utero*. Maternal dietary habits during pregnancy are thought to influence the dietary preferences of the offspring, where they are more likely to enjoy the foods that the mother has eaten (Benton 2004). This is considered to be due to the fetal recognition of odours that are present in the amniotic fluid from maternal ingestion, which may have a relationship with the acceptance of certain foods at weaning and a higher acceptance of those flavours that were present in the amniotic fluid (Schaal, Marlier and Soussignan 2000).

As flavours and odours are present in amniotic fluid and both taste and smell are developed in the fetus with the fetus swallowing amniotic fluid regularly, the first experience of flavour actually occurs prior to birth (Savage, Fisher and Birch 2007). Therefore the consumption of foods during pregnancy known to promote obesity may have a direct impact on the development of childhood obesity in the offspring due to their established food preferences. As with the presence of flavours in the fetal environment, there is a similar relationship with maternal dietary habits and breastfeeding as flavours are also present in breast milk (Savage, Fisher and Birch 2007). Infants who are breastfed seem to be more readily acceptable to try new foods when introduced (Birch 1998). Those breastfed for at least twelve months seem to be more able to regulate their own dietary intake relating to satiety than those breastfed for shorter periods (Fisher 2000). In contrast with the flavours of maternally ingested foods in breast milk, formula milk provides a consistent flavour (Savage, Fisher and Birch 2007), which may explain the acceptance of new foods as described when breastfed. Establishing healthy dietary preferences and eating patterns in early life not only have immediate nutritional benefit, but food preferences and food acceptance patterns developed in infancy are reflected in food choices made in later life (Patrick and Nicklas 2005). Therefore they may also reduce chronic disease risk when food preferences are carried into adulthood (Nicklas et al. 2001). It is also theorised that food preferences are genetic and unlearned to a certain extent. After birth infants are predisposed to accepting sweet tastes and rejecting bitter and sour flavours and this is thought to be related to the evolutionary process where the preference for sweet food will increase the consumption of energy rich foods for survival, and the aversion to bitter and sour foods will protect against the consumption of toxins (Savage, Fisher and Birch 2007).

The theory of the endogenous determination of dietary preference links with the exogenous argument for the cause of obesity being lifestyle-related, where energy intake through food and drink consumption exceeds energy usage through physical activity and leads to retained fat storage. Parental obesity is the most significant predictor for child and adulthood obesity, and overweight children are more likely to have obese parents (Parsons et al. 1999; von Kries 2002); genetic studies show that 50 to 90 per cent of adiposity in family members is explained by genetic similarity (Jacobson and Rowe 1998; Maes, Neale and Eaves 1997). Benton argues that the nature of genetic causation should be questioned, and that the behaviour patterns and extent to which those susceptible to weight gain due to the endogenous factors seek out particular environments and lifestyles (the exogenous factors) should be further explored, as little is known about the interaction with behaviour and genetic predisposition (Benton 2004). Benton questions the genetic link in isolation and argues that the family environment and the associated lifestyle behaviour may link parental and childhood obesity, and adult dietary patterns in the offspring. Food preferences established in childhood predispose dietary patterns in adulthood, and therefore the determination of obesity can be influenced by parents, as children model their food intake based on those around them. The family environment can also influence sedentary lifestyle, and the home setting is considered to be most suitable setting for prevention of childhood obesity relating to diet and physical activity (Edmunds, Waters and Elliot 2001).

The argument that obesity is caused by a combination of exogenous and endogenous factors indicates that in the case of maternal obesity, the initiation of the development of obesity potentially commences *in utero*, and that the mother and the family environment can also influence the diet and physical activity patterns after birth (Whitaker et al. 1997). There are socioeconomic inequalities pertinent to maternal obesity, the intrauterine environment, and the subsequent development of childhood obesity in the offspring. Heslehurst et al. identify that of the demographic predictors of women being obese in pregnancy, deprivation has the most significant relationship, with women living in areas of highest deprivation being almost two and a half times as likely to be obese in pregnancy than women living in areas of least deprivation (Heslehurst 2007b).

A relationship between maternal obesity and deprivation has been identified in other published research with a significantly increased frequency of low socioeconomic status in obese women when compared to women who were thin (Naeye 1990), and significantly more obese women being in receipt of Medicaid rather than having private health insurance in the U.S.A. (Rosenberg 2003). These findings reflect the findings from the Health Select Committee report, which identifies that obesity in the general population is associated with deprivation (Health Committee 2004).

Women are also more likely to smoke throughout pregnancy when they come from deprived backgrounds, increasing the fetus' intrauterine risk relating to smoking. Von Kries et al. (2002) showed that maternal nutrition is worse when women smoke during pregnancy, subjecting the infant to additional intrauterine associated risks. As obesity, smoking and poor nutrition in pregnancy is highly correlated with the inequality issues of deprivation and socioeconomic status. It is also likely that the adverse health outcomes associated with obesity in pregnancy are more prevalent in those women who have a low socioeconomic status, therefore increasing the inequalities in health for both the mother, the unborn fetus, and any offspring. Lifestyle-associated socioeconomic inequalities relate to the availability of healthy and nutritious food for the mother throughout the pregnancy, and the facilities for participating in safe physical activity.

These inequalities are further substantiated in the offspring following birth and into childhood. Children from lower social classes have poorer diets, with more frequent consumption of fatty and sugary foods than fruit and vegetables (Power and Jefferis 2002). This is potentially due to poor maternal nutrition throughout pregnancy influencing dietary preferences in the offspring as previously discussed, and childhood obesity is generally higher in offspring of women who have a lower level of education (Rodrigues et al. 1998; von Kries 2002; Moussa et al. 1994).

The cyclical nature of the development of obesity between mother and child is also apparent in the development of obesity in women. The time period between pregnancies is identified as a critical period in the development of obesity in women (Gore, Brown and Smith West 2003; Gunderson and Abrams 2000; Siega-Riz, Evenson and Dole 2004), and increasing parity in women is shown to be a significant predictor of being obese in subsequent pregnancies (Heslehurst 2007b). Therefore the very nature of reproduction has an impact on womens' subsequent development of obesity, which in turn has an impact on fertility, complications during pregnancy with health implications to both the mother and fetus, an impact on the fetal development in the uterus, early development of food preferences in the offspring, and a subsequent impact on the development of childhood obesity which has strong links with the continuation of obesity into adulthood; for female offspring this returns to the cyclical nature of obesity and the potential implications for their own reproduction.

Discussion

The interacting nature of the endogenous and exogenous determinants of obesity including genetics, the intrauterine environment, and effects of the postnatal environment make it difficult to isolate individual 'causes' of obesity for public health interventions. The life-course theory refers to the interaction and accumulation of these effects being significant, and therefore generating public health policy and effective interventions for the prevention of obesity should incorporate the interactive and multilayered determinants. Preventing childhood obesity is specifically targeted in past U.K. public health policies, such as in the 'National Healthy Schools Standard' (Department of Health and Employment 1999) and Choosing Health (Department of Health 2004a), which both address the exogenous 'lifestyle' causes of obesity. The theory behind targeting prevention of obesity during childhood is that this will reduce the prevalence of obesity in adult life, as evidence points to children who are obese becoming obese adults (Parsons et al. 1999; Whitaker et al. 1998; Garn and La Velle 1985; Unger, Kreeger and Christoffel 1990).

The 'Healthy Start' proposals for reform of the Welfare Food Scheme (Department of Health 2002b) could be considered as public health policy to prevent endogenous causes of obesity to some extent, although this is not its aim per se. Much of the policy's targets relate to diet in infant years, and therefore represent the exogenous approach of other policies. It does however highlight the need for adequate nutrition and a healthy diet throughout pregnancy and provides vouchers to help with the costs of milk, fruit and vegetables during pregnancy. In 2007, the government announced it's 'Health in Pregnancy Grant', which was an initiative to provide a lump sum payment of £190 to all women in the last months of their pregnancy (HM Treasury 2007). The intention of giving the grant was to provide women with financial support alongside advice from a health care professional at their twenty-fifth week antenatal appointment for first time mothers, and the twenty-eighth week appointment for subsequent pregnancies.

Interventions to encourage a change in lifestyle throughout pregnancy should ideally commence prior to conception in the family planning stages, or as early as possible in the pregnancy. Dieticians who run diabetes preconception clinics or infertility clinics where women attend to lose weight to conceive, report that the women are more highly motivated than under circumstances where the only outcome is weight loss (Heslehurst et al. 2007a). Women tend to be highly motivated to have healthy babies (Nankervis, Conn and Knight 2006) thus making pregnancy potentially a successful key stage in the development of initiatives for changing behaviour.

There are also initiatives to reduce smoking throughout pregnancy (Health Department Agency 2003), which are specifically aimed at disadvantaged groups. The key factors in the development of the smoking in pregnancy target are related to reducing the immediate risks of miscarriage, neonatal death, sudden infant death syndrome, and learning difficulties in offspring, and do not refer to the more long term impact of potential development of obesity in the child.

It appears that the history of policies and targets in the U.K. that have specific aims to reduce obesity levels and improve the health of the population did not take into consideration the cyclical nature of obesity, and have been primarily based on the exogenous determinants of obesity such as lifestyle initiatives to prevent the rising levels of childhood obesity. The initiatives established that could potentially prevent the endogenous factors known to promote the development of obesity were done so for alternative reasons and the potential impact on preventing the development of obesity appear to be incidental, and they are therefore not integrated in to a strategic plan. The health messages in the public domain regarding obesity relate to the associated morbidities such as heart disease and diabetes; Nankervis et al. (2006) discuss how the attention on the health burden of obesity has focussed on all-cause mortality and neglected the effects on the reproductive system and outcomes of pregnancy.

Conclusion

The arguments presented in this paper identify how the potential causes of obesity are both exogenous and endogenous, and that initiatives to attempt to reduce the rising rates of obesity aimed at lifestyle causes of childhood obesity alone may not be robust enough to be successful. The strong evidence for the endogenous causes of obesity and their relationship with exogenous causes have been virtually ignored in public health initiatives. For example the relationship between maternal dietary patterns in pregnancy, during breastfeeding, and the development of food preferences which impact on dietary habits in the offspring does not feature in any specific strategies. Future public health policy and interventions to prevent obesity should encompass all contributing factors in the development of obesity, and should be more multi-factoral and interacting between the exogenous and endogenous determinants. In particular additional focus should be instilled into the development of public health policy to address the prevention of obesity at an earlier stage in the life-cycle; *in utero*, or even prior to conception.

References

Andreasen, K.R., M.L. Andersen and A.L. Schantz. 2004. 'Obesity and Pregnancy', *Acta Obstetricia et Gynecologica Scandinavica* 83(11): 1022–29.

Barker, D.J. 1998. *Mothers, Babies and Health in Later Life.* Edinburgh: Churchill Livingstone.

Barker, D.J. 1990. 'The Fetal and Infant Origins of Adult Disease', *British Medical Journal* 301: 1111.

——— 2004. 'Role of Parents in the Determination of the Food Preferences of Children and the Development of Obesity', *International Journal of Obesity*, 28: 858–69.

Birch, L.L. 1998. 'Psychological Influences on the Childhood Diet', *Journal of Nutrition* 128: 407S–10S.

Breidahl, H.D. 1996. 'The Growth and Development of Children Born to Mothers with Diabetes', *Medical Journal of Australia* 1: 268–70.

Castro, L.C. and R.L. Avina. 2002. 'Maternal Obesity and Pregnancy Outcomes', *Current Opinion in Obstetrics and Gynecology* 14(6): 601–606.

Clark, A.M., et al. 1995. 'Weight-loss Results in Significant Improvement in Pregnancy and Ovulation Rates in Anovulatory Obese Women', *Human Reproduction* 10(10): 2705–12.

Cnattingius, S. and M. Lambe. 2002. 'Trends in Smoking and Overweight during Pregnancy: Prevalence, Risks of Pregnancy Complications and Adverse Pregnancy Outcomes', *Seminars in Perinatology* 26(4): 286–95.

———, et al. 1998. 'Prepregnancy Weight and the Risk of Adverse Pregnancy Outcomes', *New England Journal of Medicine* 338(3): 147–52.

Confidential Enquiry into Maternal and Child Health. 2004. *Why Mothers Die 2000–2002: The Sixth Report of Confidential Enquiries into Maternal Deaths in the United Kingdom.* London: RCOG Press.

——— 2007. *Perinatal Mortality 2005.* London: CEMACH.

Confidential Enquiry into Maternal Deaths. 2001. *Why Mothers Die 1997–1999.* London: RCOG Press.

Curhan, G.C., et al. 1996a. 'Birth Weight and Adult Hypertension and Obesity in Women', *Circulation* 94: 1310–15.

———, et al. 1996b. 'Birth Weight and Adult Hypertension, Diabetes Mellitus, and Obesity in U.S. Men', *Circulation* 94: 3246–50.

Department for Education and Employment. 1999. *National Healthy School Standard Guidance.* Nottingham: DfEE Publications.

Department of Health. 2002a. *Health Survey for England 2002: Health and Lifestyle Indicators for Strategic Health Authorities, 1994–2002.* London: HMSO.

——— 2002b. *Healthy Start: Proposals for Reform of the Welfare Food Scheme.* London: HMSO.

———— 2004a. *Choosing Health – Making Healthy Choices Easier*. London: HMSO.

———— 2004b. *Health Survey for England 2003*. London: HMSO.

———— 2004c. *National Service Framework for Children, Young People and Maternity Services*. London: HMSO.

———— 2005. *Health Survey for England 2004*. London: HMSO.

Dorner, G. and A. Plagemann. 1994. 'Perinatal Hyperinsulinism as Possible Predisposing Factor for Diabetes Mellitus, Obesity and Enhanced Cardiovascular Risk in Later Life', *Hormone and Metabolic Research* 26: 213–21.

Edmunds, L., E. Waters and E.J. Elliot. 2001. 'Evidence-based Management of Childhood Obesity', *British Medical Journal* 323: 916–19.

Fisher, J.O. 2000. 'Breast Feeding through the First Year Predicts Maternal Control in Feeding and Subsequent Toddler Energy Intakes', *Journal of the American Dietetic Association* 100(6): 641–46.

Foresight. 2007. *Tackling Obesities: Future Choices – Project Report*. London: Department of Innovation, Universities and Skills, Government Office for Science.

Garn, S.M. and M. La Velle. 1985. 'Two Decades Follow-up of Fatness in Early Childhood', *American Journal of Diseases of Children* 139: 181–85.

Godfrey, K.M., et al. 1997. 'Maternal Birthweight and Diet in Pregnancy in Relation to the Infant's Thinness at Birth', *BJOG: An International Journal of Obstetrics and Gynaecology* 104: 663–67.

Gore, S.A., D.M. Brown and D. Smith West. 2003. 'The role of postpartum weight retention in obesity among women: A review of the evidence', *Annals of Behavioral Medicine* 26(2): 149–59.

Guelinckx, I., et al. 2008. 'Maternal Obesity: Pregnancy Complications, Gestational Weight Gain and Nutrition', *Obesity Reviews* 9: 140–50.

Gunderson, E.P., and B. Abrams. 2000. 'Epidemiology of Gestational Weight Gain and Body Weight Changes after Pregnancy', *Epidemiologic Reviews* 22(2): 261–74.

Health Committee. 2004. *Obesity – Third Report of Session 2003–2004*, vol. 1, HC 23.1. London: HMSO.

Health Development Agency. 2003. *Meeting Department of Health Smoking Cessation Targets*. London: HDA.

Heslehurst, N., et al. 2007a. 'Obesity in Pregnancy: A Study of the Impact of Maternal Obesity on NHS Maternity Services', *BJOG: An International Journal of Obstetrics and Gynaecology* 114: 334–42.

————, et al. 2007b. 'Trends in Maternal Obesity Incidence Rates, Demographic Predictors, and Health Inequalities in 36,821 Women over a 15-Year Period', *BJOG: An International Journal of Obstetrics and Gynaecology* 114: 187–94.

————, et al. 2008. 'The impact of maternal BMI status on pregnancy outcomes with immediate short-term obstetric resource implications: a meta-analysis', *Obesity Review* 9: 635–83.

HM Treasury. 2007. *The Health in Pregnancy Grant (Entitlement, Amount and Administration) Regulations 2007*. London: HM Treasury.

Institute of Medicine. 1990. *Nutrition during Pregnancy, Subcommittee on Nutritional Status and Weight Gain during Pregnancy*. Washington, DC: National Academic Press.

———— 2009. Weight Gain during Pregnancy: Reexamining the Guidelines. National Academic Press, Washington DC: National Academic Press.

Jacobson, K.C. and D.C. Rowe. 1998. 'Genetic and Shared Environment Influences on BMI: Interactions with Race and Sex', *Behavioral Genetics* 337: 869–73.

Kanagalingam, M.G., et al. 2005. 'Changes in Booking Body Mass Index over a Decade: Retrospective Analysis from a Glasgow Maternity Hospital', *BJOG: An International Journal of Obstetrics and Gynaecology* 112: p. 1431–33.

Larsen, C.E., M.K. Serdula and K.M. Sullivan 1990. 'Macrosomia: Influence of Maternal Overweight among a Low-income Population', *American Journal Obstetrics Gynecology* 162(2): 490–94.

Lashen, H., K. Fear and D.W. Sturdee. 2004. 'Obesity is Associated with Increased Risk of First Trimester and Recurrent Miscarriage: Matched Case-control Study', *Human Reproduction* 19(7): 1644–46.

Lawlor, D.A. and N. Chaturvedi. 2006. 'Treatment and Prevention of Obesity – Are There Critical Periods for Intervention?', *International Journal of Epidemiology* 35: 3–9.

Lewis, G. (ed.). 2007. *The Confidential Enquiry into Maternal and Child Health (CEMACH). Saving Mothers' Lives: Reviewing Maternal Deaths to Make Motherhood Safer 2003–2005. The Seventh Report on Confidential Enquiries into Maternal Deaths in the U.K.* London: CEMACH.

Linne, Y. 2004. 'Effects of Obesity on Women's Reproduction and Complications during Pregnancy', *Obesity Reviews* 5: 137–43.

Maes, H.H., M.C. Neale and L.J. Eaves. 1997. 'Genetic and Environmental Factors in Relative Body Weight and Human Adiposy', *Behavioral Genetics* 28: 265–78.

Morin, K.H. 1998. 'Perinatal Outcomes of Obese Women: A Review of the Literature', *Journal of Obstetric, Gynecologic, and Neonatal Nursing* 27(4): 431–40.

Moussa, M.A., et al. 1994. 'Factors Associated with Obesity in School Children', *International Journal of Obesity* 18: 513–15.

Naeye, R.L. 1990. 'Maternal Body Weight and Pregnancy Outcome', *American Journal of Clinical Nutrition* 52: 273–79.

Nankervis, A.J., J.J. Conn and R.L. Knight. 2006. 'Obesity and Reproductive Health', *Medical Journal of Australia* 184(2): 51.

Nicklas, T.A., et al. 2001. 'Family and Child-care Provider Influences on Preschool Children's Fruit, Juice, and Vegetable Consumption', *Nutrition Reviews* 59(7): 224–35.

Norman, R.J. and A.M. Clark. 1998. 'Obesity and Reproductive Disorders: A Review', *Reproductive Fertility Development* 10: 55–63.

Parsons, T.J., et al. 1999. 'Childhood Predictors of Adult Obesity: A Systematic Review', [Review; 283 references] *International Journal of Obesity and Related Metabolic Disorders* 23(12): S1–S107.

Patrick, H. and T.A. Nicklas. 2005. 'A Review of Family and Social Determinants of Children's Eating Patterns and Diet Quality', *Journal of the American College of Nutrition* 24(2): 83–92.

Pettigrew, R. and D. Hamilton-Fairley. 1997. 'Obesity and Female Reproductive Function', *British Medical Bulletin* 53(2): 341.

Pettitt, D.J., et al. 1993. 'Diabetes and Obesity in the Offspring of Pima Indian Women with Diabetes during Pregnancy', *Diabetes Care* 16(1): 310–14.

Plaguemann, A., et al. 1997. 'Glucose Tolerance and Insulin Secretion in Children of Mothers with Pregestational IDDM or Gestational Diabetes', *Diabetologia* 40(9): 1094–1100.

Poston, L., and P. Taylor. 2007. 'Obesity in Pregnancy and the Next Generation: What Can We Learn from Animal Models?' in P. Baker et al. (eds), *Obesity and Reproductive Health*. London: RCOG Press.

Power, C. and B.J. Jefferis. 2002. 'Fetal Environment and Subsequent Obesity: A Study of Maternal Smoking', *International Journal of Epidemiology* 31(2): 413.

———, et al. 2003. 'Combination of Low Birth Weight and High Adult Body Mass Index: At What Age Is It Established and What Are Its Determinants?', *Journal of Epidemiology and Community Health*, 2003. 57(12): 969.

Reilly, J.J. and Z.C. McDowell. 2003. 'Physical Activity Interventions in the Prevention and Treatment of Paediatric Obesity: Systematic Review and Critical Appraisal', *Proceedings of the Nutrition Society* 62: 611–19.

Rodrigues, S., et al. 1998. 'Obesity among Offspring of Women with Type 1 Diabetes', *Clinical and Investigative Medicine – Médecine Clinique et Expérimentale* 21(6): 258.

Rosenberg, T.J., et al. 2003. 'Prepregnancy Weight and Adverse Perinatal Outcomes in an Ethnically Diverse Population', *Obstetrics and Gynecology* 102(5): 1022–27.

Savage, J.S., J.O. Fisher and L.L. Birch. 2007. 'Parental Influence on Eating Behaviour: Conception to Adolescence', *Journal of Law, Medicine and Ethics* 35(1): 22–34.

Schaal, B., L. Marlier and R. Soussignan. 2000. 'Human Foetuses Learn Odours from Their Pregnant Mothers' Diet', *Chemical Senses* 25: 729–37.

Siega-Riz, A.M., K.R. Evenson and N. Dole. 2004. 'Pregnancy-related Weight Gain – A Link to Obesity?', *Nutrition Reviews* 62(7): S105–111.

Silverman, B.L., 1991. 'Long-term Prospective Evaluation of Offspring of Diabetic Mothers', *Diabetes* 40(2): 121–25.

Unger, R., L. Kreeger and K.K. Christoffel. 1990. 'Childhood Obesity: Medical and Familial Correlates and Age of Onset', *Clinical Paediatrics* 29: 368–73.

Van der Meulin, J. 2002. 'Maternal Smoking during Pregnancy and Obesity in the Offspring', *International Journal of Epidemiological Association* 31: p. 420–21.

Von Kries, R., et al. 2002. 'Maternal Smoking during Pregnancy and the Risk of Childhood Obesity', *American Journal of Epidemiology* 156(10): 954–61.

Waller, K., et al. 2007. 'Prepregnancy Obesity as a Risk Factor for Structural Birth Defects', *Archives of Pediatrics and Adolescent Medicine* 161(8): 745–50.

Wang, J.X., M.J. Davies and R.J. Norman. 2002. 'Obesity Increases the Risk of Spontaneous Abortion during Infertility Treatment', *Obesity Research* 10(6): 551–54.

Whitaker, R.C., et al. 1997. 'Predicting Obesity in Young Adulthood from Childhood and Parental Obesity', *New England Journal of Medicine* 337: 869–73.

———, et al. 1998. 'Early Adiposity Rebound and the Risk of Adult Obesity', *Pediatrics* 101(3): E5.

Chapter 7

CULTURE, DIET AND
THE MATERNAL BODY:
GHANAIAN WOMEN'S PERSPECTIVES
ON FOOD, FAT AND CHILDBEARING

Ama de-Graft Aikins

Introduction

Obesity is a significant problem for African women. In some countries, more than half of the women are overweight or obese. Female obesity rates are greater than male rates. In the late 1990s obesity was seventeen times as common in Gambian women as in men, four times as common in Moroccan women as in men, and three times as common in South African women (Prentice 2006). Obesity is a risk factor for a number of chronic diseases including diabetes, cardiovascular diseases and cancers, which are assuming public health significance in many African countries (BMJ 2005; WHO 2005; WHO/FAO 2005).

Africa's obesity burden, like the global obesity pandemic, is driven by a 'nutrition transition' – a coexistence of under-nutrition and over-nutrition – rooted in geo-political and sociocultural factors. Popkin (1998 2005), identifies powerful external factors such as changes in global food supply, relative low cost of foodstuffs in particular highly refined oils and carbohydrates, the promotion of increased use of motorized transport and energy-saving devices, and increasingly sedentary employments that create 'obesogenic' environments. These processes and products of globalisation are compounded by sociocultural changes in dietary patterns and physical activity.

Researchers observe that sociocultural and behavioural factors contribute significantly to the disproportionate prevalence of obesity among African women. The association of fat with beauty, status and wealth in many African countries is often heavily implicated. In recent years, the reification of fat has been reinforced by the association between weight loss, extreme thinness and HIV/AIDS (Kruger et al. 2005). Institutionalised cultural practices and women's internalisation of cultural norms may also predispose them to gaining weight at significant life stages in particular at the start of marriage (Maletnlema 2002), during childbirth, and during middle age (Siervo et al. 2005).

While there is increasing regional research on female obesity, most studies do not explore the link between obesity risk and significant life stages. For example while anthropological and demographic evidence suggests that the childbearing phase of an African woman's life may present weight gain and obesity risks, existing studies either exclude pregnant and nursing mothers or do not probe the link between childbearing, food practices and obesity. Research in the Western context suggests a strong correlation between multiple childbirths and sustained weight gain. Obesity during pregnancy increases maternal health risks including chronic conditions such as hypertension and diabetes (Dixit and Girling 2008). Thus focusing on this period of women's lives may yield important insights for obesity prevention and general health interventions for African women.

In this chapter, I report a social-psychological study that explores the link between childbearing and obesity among Ghanaian women. Generally, the study aimed to explore the extent to which the childbearing life-stage presents obesity risk for Ghanaian women and what this means for women's diet, weight and health management strategies during and after childbearing. Three research questions guided the study:

(1) How does culture structure experiences of childbearing, in particular experiences and practices that present obesity risk for women, such as food consumption and physical activity?
(2) How do women respond to these cultural factors during the childbearing period and what shapes these responses?

(3) What are the implications of (1) and (2) for diet, weight and health management interventions for women?

Obesity among Ghanaian women: available evidence

Ghana's obesity epidemic has been attributed to urbanization, globalisation, affluence and changing lifestyles (in particular sedentary occupations and consumption of a wider diversity of local and foreign foods) (Agyei-Mensah and de-Graft Aikins 2007; Amoah 2003; Levine et al. 1999). These local discussions are aligned to global perspectives on the impact of 'obesogenic' environments on population weight gain (Popkin 2005; WHO/FAO 2003).

In Ghana, as in much of Africa, obesogenic environments impact differently for men and women. In a 2003 national obesity survey obesity was almost three times as common in women as in men (Britwum et al. 2005). Demographic data shows accelerated increase in obesity prevalence among women over the last twenty years. Obesity rates doubled from 10 per cent in 1993 to 25.3 per cent in 2003 (GSS et al. 2004). Rates further increased 1.2-fold to 30.5 per cent in 2008 (Dake 2009). Across the country – for each period of data collection – highest levels of overweight and obesity occurred among urban, educated, high income and married women aged between thirty and fifty. There is an established relationship between adult female obesity and chronic diseases such as hypertension and diabetes in urban Ghana (Hill et al. 2005). Hypertension is a leading cause of maternal mortality in the country (Kumi-Aboagye 2008; Lassey and Wilson 1998).

While sociocultural factors are implicated, the available evidence is complex and nuanced. In regional accounts strong, and generalised, associations are made between reification of fat and obesity among women. For instance, Prentice (2006) asserts that while Western countries tend to stigmatise fat, this 'psychological break' is 'absent' in African societies. The available Ghanaian data suggests that this 'psychological break' may be absent in some regions, but is present in others. In the Ashanti region, for example, fat is reified culturally and this influences women's weight management practices. Young and middle-aged women in Kumasi, Ghana's second largest city, like middle-aged women in the Gambia (Prentice 2006), take over-the-counter medications to increase their weight (O. Boateng, personal communication 2006). The influence of culture in relation to body image and size appears to be strong for women in other regions, such as the Greater Accra Region. Studies show that weight management strategies are influenced by cultural norms (Anum and de-Graft Aikins forthcoming) and the wishes of women's significant others, especially their husbands (Duda et al. 2006). However, research suggests that for some women in the Greater Accra region 'interest in healthy living outweighs presumed cultural norms for obesity' (Duda et al. 2006). Few studies have explored the relationship between obesity risk and women's life-stages.

Conceptual framework

The available literature on obesity among Ghanaian women provides three important starting points for this study. First, the epidemiological data on high-risk groups for overweight and obesity suggest marriage and childbearing life stages as potential obesogenic periods. Second, like other African countries, culture exerts a strong influence on women to gain weight in order to fit shared perceptions of beauty, status and wealth. Third, women may respond to this cultural pressure at different life stages in a variety of ways, ranging from internalisation to resistance. To explore these ideas, the study draws on social representations theory (SRT), and its central concept 'cognitive polyphasia'. SRT employs the classical social psychology approach to knowledge production and use that situates its level of analysis at the interface of the individual and society (cf. Bartlett 1932; Moscovici and Duveen 2000). It aims to 'conceptualise, simultaneously, both the power of society and the agency of individuals' (Gervais et al. 1999: 422).

SRT's central concept of 'cognitive polyphasia' has been defined as 'the dynamic coexistence ... of distinct modalities of knowledge, corresponding to defined relationships between human beings and their surroundings' (Moscovici 1961/1976: 186). The concept draws attention to the ways groups and individuals draw eclectically, and often in opposing ways, on different modalities or 'stocks' of knowledge – e.g., cultural, scientific, religious and common sense knowledge – in everyday life (Flick 1998). By focusing on the interface between sociocultural and individual processes and on the complex dimensions of knowledge production and use, SRT allows systematic examination of the sources, contents and functions of the knowledge women draw on prior to and during childbearing. By placing emphasis on subjective and intersubjective experiences, SRT also facilitates understanding of the complexity of everyday experience and allows the implications of women's use of knowledge on specific health outcomes to be identified.

Study informants and methods

Thirty-five women of different ethnicities (Akan, Dagao, Ewe, Ga and Hausa) and age groups (twenty-seven to seventy-five) were recruited for the study. All except two women lived in Accra. Two lived outside Accra in Takoradi (Western region) and Akosombo (Eastern region) but had lived in Accra previously where they had given birth to and raised some of their children. Two had had all their children abroad, in the U.K. and in Canada, but maintained Ghanaian food practices during their pregnancies. Like many urban Ghanaians, whether transient migrants or permanent residents, these women maintained strong psychological ties with their ethnic groups if not rural hometowns (Acquah 1958; Busia 1950; Allman and Tashjian 2000).[1] Table 7.1 presents the profile of the study women.

With the exception of one, all informants were educated. Income status varied from low-income (small-scale traders) to high income (legal and academic consultants). While some women noted that food budgets increased during pregnancy, none of the women reported experiencing food insecurities (in terms of physical and financial access) during their pregnancies. Most women were married at the time of first pregnancy and remained married to the fathers of their children at the time of interviews. One woman was divorced; one was single. The number of children women had ranged between one (for the youngest informants) and four. The youngest age for a first birth was eighteen (for a woman in her mid-twenties); the oldest age for a first birth was thirty-six (a woman who was breastfeeding at the time of her interview in August 2007). One woman was pregnant at the time of her interview; two were breastfeeding.

After gathering demographic information (age, ethnicity, education, occupation(s),[2] number and ages of children), the women were asked questions on:

(1) ante-natal histories (probing health and well-being issues and the sociocultural context of childbearing);
(2) recommended and prohibited foods for pregnancy and breastfeeding (including sources of knowledge and functions of the foods);

Table 7.1: *Participant profiles*

	Number of participants
Age	
21 – 40	13
41 – 60	14
61 – 80	8
Ethnicity	
Akan (Akwapim, Asante, Fante)	15
Dagao	5
Ewe	5
Ga	6
Hausa	4
Education	
None	1
Primary	6
Secondary	7
Tertiary (polytechnic, vocational training)	5
University	16
Occupation	
Professional (law, management, academic)	9
Professional (nursing, teaching, secretarial)	6
Trader (market, store, street)	9
Retired (but working: consulting, running	3
kindergarten, etc.)	1
Retired (no work)	1
Unemployed	6
Other (e.g., traditional leader)	

(3) food practices during their pregnancies and breastfeeding; and

(4) perspectives on food fat and childbearing and the relationship between these perspectives and weight management practices during and after childbearing.

All women generated their pregnancy food list using a free recall approach. Question (2) produced a general pregnancy and breastfeeding food list (linked to cultural and other social sources), question (3) a respondent-specific pregnancy and breastfeeding food list (linked to individual experiences). Sources of food knowledge were explored, as well as functions of foods, including their fattening properties.

As interviews progressed and perspectives approached 'meaning saturation',[3] items on the cumulative pregnancy food list were used as memory prompts (for example, I spoke to some women who said that snails were good for pregnancy: 'what's your opinion on that?'). A key limitation of this open-ended approach with a cohort of women, some of whom were required to remember their pregnancy experiences dating back a number of years or decades, was that certain details of everyday pregnancy experience such as food portions, quantity of herbs used in cooking, frequency of food intake, and amount of weight gain were lost. This kind of detailed information is best gathered by direct observation or the use of food diaries (Messer 1989) and would necessitate studying pregnant and lactating women over extended periods of pregnancy and breastfeeding. However, asking women to speak about their pregnancy experiences brought, for many, vivid memories of the physical and physiological aspects of their pregnancies.

Many began their accounts of pregnancy experience with detailed information on bodily changes: these bodily changes were strongly linked to food consumption (for example, hypersensitivity to smell and inability to eat spicy foods; uncontrolled craving for 'out of the ordinary' foods). Thus, acute memories of bodily changes facilitated memory of food practices. In later interviews, probes on the physical experience of pregnancy were used alongside food item prompts to maximise recollection of food practices. I interviewed twenty-four Akan and Ga women. The remaining women were interviewed by two male postgraduate research assistants, one with a bachelor's degree in nutrition. We recruited our respondents through our multi-ethnic social and professional networks. We used a mix of languages in interviewing: English, Fante, Asante-Twi and Dagao. Informed consent was sought from all respondents. Notes were taken for all interviews;[4] and some women were interviewed more than once.

The data was analysed using the technical definition of cognitive polyphasia as a conceptual tool. This involved first identifying consensus, conflict and absence within and across responses to the four interview areas empirical for each interview, then comparing the set of interviews with the same criteria. The Consensus-Conflict-Absence approach was developed within the context of analysing large-scale multilingual qualitative data on rural and urban Ghanaian diabetes experiences (for further discussion, see de-Graft Aikins 2004, 2005).

Study findings

Results will be presented in four sections:

(1) sources and modalities of food knowledge;
(2) contents of food knowledge and functions of foods;
(3) food practices during pregnancy and breastfeeding and mediating factors;
(4) weight management during pregnancy and breastfeeding

1. Sources and modalities of food knowledge

Women drew their knowledge of pregnancy and breastfeeding foods from five sources: (1) family and friends;' (2) school and university; (3) the biomedical sphere (doctors, nurses); (4) media (newspapers, radio and television); and (5) unique pregnancy experiences (bodily demands).

Knowledge drawn from family and friends – hereafter, lifeworld[5] – can be categorised as cultural knowledge. The knowledge was handed down from generations and through mothers and other female members of the extended family. This knowledge encompassed physiological, physical as well as intersubjective and sociocultural dimensions of pregnancy experiences and as such instructed a broad repertoire of food and health practices. It was ethnically bound, so that foods specific to certain groups and regions of Ghana appeared on the list of women originally from these regions, however long they had lived in Accra. All women cited their lifeworld as a primary source of knowledge for managing pregnancy and breastfeeding experiences.

School/university knowledge and knowledge from biomedical sphere constituted expert biomedical knowledge. This can be described as scientific knowledge: highly technical knowledge based on nutritional science and obstetrics that focuses on a restricted repertoire of practical nutrition management routines aimed at addressing the physical and physiological dimensions of pregnancy. The biomedical sphere was cited by more women as an important source of knowledge, compared to school and/or university. The latter was cited by two groups of women: northern women who received life-skills teaching at their northern region schools, including reproductive health lessons; and women with careers in the biomedical sciences who had attended lectures on reproductive health and maternal nutrition. Women appeared to draw on scientific knowledge only during the pregnancy period.

The mass media was cited as a source of knowledge by six women. The contents of information were not explicitly outlined: women made general references to reading about general reproductive health matters in the newspapers or learning about these matters on radio and television. Generally Ghanaian media health reports draw largely on biomedical information, either commissioning medical and health experts to write newspaper articles and discuss issues for radio or television audiences or culling articles from local magazines and international newspapers. The knowledge may therefore be described as 'scientised knowledge': its contents often have a core basis in scientific knowledge but may be subjected to lay revisions and simplifications that distort meaning.

Body-self knowledge has been defined as subjective knowledge of one's unique state of physical and psychological balance (Helman 2000: 14): the term corresponds loosely with references within sociocultural theories of emotion to 'embodied knowledge' (cf. Lupton 1998). Body-self knowledge has been reported to mediate health maintenance and illness management practices in a variety of contexts (Angel and Guarnaccia 1989; Bates et al. 1997) including Ghana (de-Graft Aikins 2004). Body-self knowledge emerged as an important source of knowledge during actual experiences of pregnancy and breastfeeding. This knowledge modality was particularly important for women with more than one child, because body-self knowledge shaped food practices throughout all pregnancies and to a large extent mediated use of other knowledge modalities.

2. Contents of food knowledge and functions of foods
Recommended foods for pregnancy and breastfeeding

The women generated a wide range of recommended foods and food supplements for pregnancy and breastfeeding. The majority of foods were indigenous to Ghanaian diets generally (for example green leafy vegetables, legumes, meat and fish, fruits). Meals were ethnically bound. With respect to these ethnically bound food practices some notable – but expected – differences occurred. There were more references to fish in southern diets than northern; more references to daily use of leafy green vegetables in northern diets compared to southern; and the mode of preparation of carbohydrate-based meals differed between southern and northern women (southern women ate fermented maize/millet dough-based staples, northern women ate non-fermented maize/millet dough-based staples). These differences have nutritional implications. Fish-based diets, vegetable-based diets and fermented maize/millet staples are more nutritious and healthy. The indigenous foods listed by women are termed traditional foods and supplements: their indigenous status allows identification of culturally mediated food consumption patterns. The food list also included foreign foods such as beverages (tea, 'Milo', 'Ovaltine', 'Horlicks'), dairy products (milk, yoghurt) and biscuits (digestives, cream crackers). This list cut across ethnicity.

All women noted that the food list they generated for pregnancy applied to breastfeeding. A key difference, however, was the emphasis their mothers (and/or other important female members of their extended families) placed on soups and 'fluids'. Palm-nut soup emerged as an important breastfeeding food for all women. Other soups were ethnically bound: for example, *agushi* (melon seeds) soup for northern women. Common 'fluids' mentioned by the majority of women were porridge (*hausa koko* (cornmeal porridge with spices), rice water, millet porridge), 'mashed kenkey' (a traditional 'shake' or 'smoothie' made with mashed kenkey – a corn dough meal – blended with water, milk and sugar), and 'Milo'.[6]

The contents of the pregnancy and breastfeeding food lists, both traditional and foreign, were similar across different age groups. That is to say pregnancy and breastfeeding food beliefs and knowledge had not changed much across three generations (between women in their late sixties, forties and twenties).

Prohibited foods for pregnancy and breastfeeding

Some women made reference to prohibited foods for pregnancy. There was general agreement that excess fat and sugar was bad for pregnancy. This knowledge appeared to be influenced by antenatal information provided by health professionals. Only one specific health-provider-related prohibited food was mentioned. One woman noted that her doctor advised her to stop eating bread as it blocked absorption of essential vitamins and minerals.

Beyond this, references to specific prohibited foods were culturally driven. Some Akwapim women stressed that groundnut soup caused heartburn and was therefore prohibited in pregnancy. One Fante woman placed extremely sweet fruits – sugarcane, pineapple, coconut – under the prohibited foods list. She had been informed that sugarcane had abortive functions and had to be restricted during pregnancy. Another Fante woman had been told by women in her family that snails were bad for pregnancy: they caused babies to drool excessively. She did not believe this herself.

The status of other prohibited foods was not as clear-cut. For some women eggs were prohibited in pregnancy: for others eggs were recommended. A final category of prohibited 'foods' included items such as clay and ice: their consumption by pregnant women in a variety of settings has been attributed to 'pica'.[7] Because these items belong to the participant-specific pregnancy food list, (i.e., food that is craved and/or consumed during pregnancy but may not be generally recommended) I discuss it within the context of consumption in the next section.

The list of prohibited foods for breastfeeding was shorter than the pregnancy list. In order of dominance the following foods were mentioned: oily foods (four women), okro (one woman), sugar (one woman) and tea (one woman). Of these, only okro was attributed specific negative functions: it was believed to cause a baby 'to have a running stomach'.[8] It was difficult to make strong ethnic attributions to these prohibited breastfeeding foods as they were minority views. Figure 7.1 presents the list of prohibited foods for pregnancy and breastfeeding.

Functions of (recommended) pregnancy and breastfeeding foods

Most women attributed three dominant nutritional functions to pregnancy foods. Some foods, such as green leafy vegetables (consumed as stews or soups), were understood to 'give blood' (or prevent anaemia). Others, such as palm nut (consumed as soups), were understood to 'give body'. Often, the term 'to give body' was used interchangeably to mean: (1) strengthening the body; and (2) fattening the body. Finally, herbs and other food supplements such as *prekese (Tetrapleura tetraptera)*, *alefi*, *dawadawa* and *kwawu nsua* were central to meals and were believed to aid the processes of giving blood and body as well as minimising physiological disruptions like nausea.

Breastfeeding foods were attributed one dominant function: to maximise breastmilk production. All the soups and fluids mentioned specifically by women were attributed with this function. Most women noted that these breastfeeding foods, especially palm-nut soup, were fattening. Table 7.2 outline functions attributed to pregnancy foods.

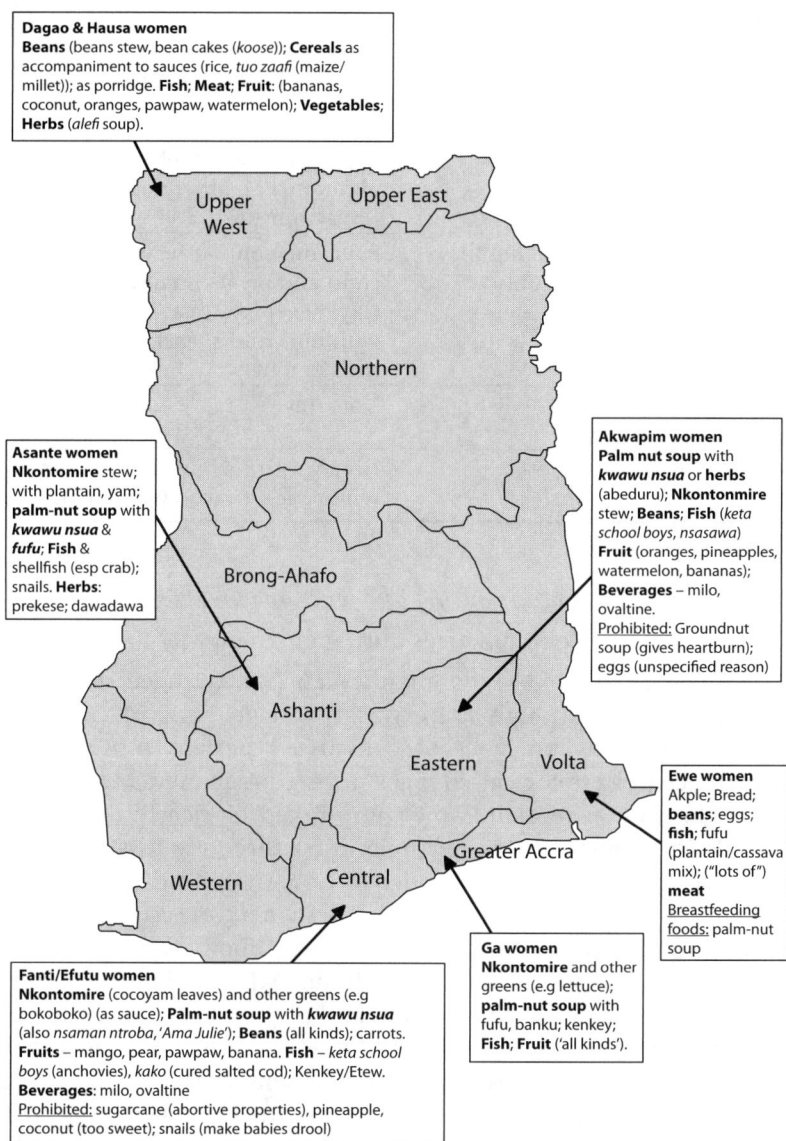

Dagao & Hausa women
Beans (beans stew, bean cakes (*koose*)); **Cereals** as accompaniment to sauces (rice, *tuo zaafi* (maize/millet)); as porridge. **Fish**; **Meat**; **Fruit**: (bananas, coconut, oranges, pawpaw, watermelon); **Vegetables**; **Herbs** (*alefi* soup).

Upper West

Upper East

Northern

Asante women
Nkontomire stew; with plantain, yam; **palm-nut soup** with *kwawu nsua* & *fufu*; **Fish** & shellfish (esp crab); snails. **Herbs**: prekese; dawadawa

Brong-Ahafo

Ashanti

Akwapim women
Palm nut soup with *kwawu nsua* or **herbs** (abeduru); **Nkontonmire** stew; **Beans**; **Fish** (*keta school boys, nsasawa*) **Fruit** (oranges, pineapples, watermelon, bananas); **Beverages** – milo, ovaltine.
Prohibited: Groundnut soup (gives heartburn); eggs (unspecified reason)

Eastern

Volta

Ewe women
Akple; Bread; **beans**; eggs; **fish**; fufu (plantain/cassava mix); ("lots of") **meat**
Breastfeeding foods: palm-nut soup

Greater Accra

Western

Central

Fanti/Efutu women
Nkontomire (cocoyam leaves) and other greens (e.g bokoboko) (as sauce); **Palm-nut soup** with *kwawu nsua* (also *nsaman ntroba,'Ama Julie'*); **Beans** (all kinds); carrots. **Fruits** – mango, pear, pawpaw, banana. **Fish** – *keta school boys* (anchovies), *kako* (cured salted cod); Kenkey/Etew. **Beverages**: milo, ovaltine
Prohibited: sugarcane (abortive properties), pineapple, coconut (too sweet); snails (make babies drool)

Ga women
Nkontomire and other greens (e.g lettuce); **palm-nut soup** with fufu, banku; kenkey; **Fish**; **Fruit** ('all kinds').

Note: Breastfeeding foods that overlap with pregnancy foods are shaded. Breast-feeding foods that are mentioned apart from pregnancy foods are presented under a separate heading. *Kwawu nsua/nsaman troba*: small vegetables from the eggplant family; *Kenkey*: Akan/Ga staple made from fermented corn dough.

Figure 7.1: *Pregnancy and breastfeeding foods by ethnic group*

Table 7.2: *Functions of pregnancy foods and food supplements*

Functional categories of pregnancy foods	Food groups
Preventing anaemia ('giving blood')	Leafy green vegetables (e.g., nkontomire); other vegetables (kwawu nsua)
Strengthening and fattening the body and increasing fertility ('giving body')	Palm-nut soup; fish, shellfish (esp. crab) and meats; snails; blood tonics; clay; beverages (tea, Milo, Ovaltine with milk and sugar); food supplements
Minimising physiological disruption (nausea, vomiting)	Kola nuts
Maximising health of fetus	Honey, carrots (give a baby 'very white eyes' (Dagao woman, 32))

3. Food practices during pregnancy and breastfeeding
Food practices during pregnancy

Pregnancy, as one elderly woman (Akan (Efutu), aged sixty-nine) described, felt as though there was 'a foreign body in your body making demands on you'. As discussed earlier, most women experienced physiological changes during pregnancy. Memories of these changes were acute and were linked to clearly demarcated periods in pregnancy (for example, 'first three months', 'first five months', 'final month') or gender of unborn child (it felt different carrying boys versus girls for some). These memories constituted the lens through which relationships with food were discussed. Pregnancy was a period during which the majority of women gained intimate knowledge of their bodies and physiological processes. For those who had more than one child, the first pregnancy provided a sensory map for future pregnancies: some sensed they were pregnant before conclusive medical tests because their bodies underwent clear predictable changes; others knew they were expecting either girls or boys.

For many women these physiological changes were primary mediators of food consumption during pregnancy. That is, depending on the severity of physiological changes one could adhere to a well-balanced nutritious diet or not. Physiological changes mediated food consumption in three main ways, corresponding to three common experiences described by the women.

Hypersensitivity and Hyperemesis gravidarum (HG)

About half of the women had experienced hypersensitivity to smell, light and motion during the first trimester. Hypersensitivity to smell was linked to nausea, spitting and vomiting. Women who experienced nausea in their first trimester were therefore unable to eat a balanced diet as the smell of cooked (especially fried) and/or spicy foods exacerbated nausea. Most listed a restricted range of simple – 'clean' (raw, steamed, boiled, roasted) – foods they could consume during this period (see Table 7.3). During this period food supplements provided by antenatal clinics become central to maternal health. Four women reported experiences and symptoms that suggested Hyperemesis gravidarum (HG), a complication of pregnancy that affects 1 per cent of women. These women had extreme morning sickness with persistent nausea and vomiting and suffered weight loss in the first trimester.

Increased appetite and craving out-of-ordinary foods

Some women experienced increased appetite throughout their pregnancy or after the first difficult three months. During periods of increased appetite most women ate all the recommended foods, often several times a day. For example, one first-time lactating mother (Ga, aged thirty-six) recalled experiencing increased appetite after the first three months of (nausea-induced) limited eating. She

Table 7.3: *Foods consumed during the first trimester*

Raw foods	Cooked foods	Dry/processed foods
'fruits: oranges, pineapples, watermelon and food supplements' (44, Akwapim)	'kobia, kpakpo shitob, okro (steamed) ... kola nuts' (54, Ga) 'pepper, fish (either fried or smoked), hot banku. Light soup. Tea with milk.' (69, Ga) 'fresh pepper, steamed crab and 'kenkey'; light soup.' (47, Fante) 'only dried fish' (75, Asante) 'rice and bread' (69, Fante)	'Plain tea (without milk or sugar)' (64, Ga) 'cereals ... cocoa' (65, Krobo) Ice cubes and food supplements (37, Asante; 47, Akwapim). 'Soya milk, digestive biscuits and food supplements' (36, Ga)

ª cured and salted tilapia
ᵇ small green aromatic chilli peppers

ate heavy starch and protein-based foods on average seven times a day and at unstructured times, putting on 3 kgs of weight. Her doctor advised a 'weight check'. Weight gain was common during periods of increased appetite (see the next section of this chapter) and posed health risks to the mother and unborn child.

Part of the experience of increased appetite was sudden and often prolonged craving for 'out of the ordinary' foods. These 'out-of-ordinary' foods tended to be largely foreign (hamburgers, apples) or foods from different ethnic groups (e.g., an elderly Fante woman who craved Hausa porridge). A minority of women (across the age spectrum) craved and consumed pica: Akwapim clay, chalk and ice were three such substances. The majority of foods belonged to the general list of recommended pregnancy foods. Of the remainder, one category – the 'non-food' items clay, chalk and ice – were not recommended within lifeworlds or by healthcare providers. As one elderly Fante women noted: 'I used to eat clay, Akwapim clay. My body craved it. Nobody told us to eat it, we just wanted it.'

The second category constituted foods such as ice cream and hamburgers and drinks such as 'Malta Guinness' that, by virtue of their high-fat and high-sugar content, belonged to the prohibited food list. Craved foods served different purposes for different women: nutritional (in the example given above calcium boost in Akwapim clay), physiological (the nausea-reducing properties of *bisi* (kola nut) for one middle-aged Ga woman) and psychological (the comfort provided by the smell, texture and taste of *akrantie* for the elderly Fante woman).

Pregnancy-induced illness

The lone observation by one elderly Fante woman that 'nyinsan bi ye yariba' (some pregnancies are illnesses) was true for about a third of study participants. Four women (aged thirty-seven, forty-five, forty-six, fifty) developed hypertension during their pregnancies, two women had malaria in their first trimester, three women experienced persistent heartburn when they ate spicy foods, one woman (aged forty-four) developed pre-eclampsia during her second pregnancy, and an elderly Ga woman spent the whole duration of all of her four pregnancies in hospital on a vitamin-enhanced drip. Most women in this sub-group were placed on strict – low salt, low fat, no spice – diet restrictions by their doctors during their pregnancies. These diet restrictions were often difficult to maintain for all the women: struggling with weight loss strategies was a common experience. The remaining number of women reported having had 'healthy pregnancies', including those who had hypersensitivity issues in their first trimester.

What becomes apparent from these accounts of pregnancy experiences is that bodily processes, and by extension the pregnant body, constituted an important legitimate source of maternal nutrition knowledge. As the elderly Ga women who spent all her four pregnancies in hospital on a drip observed: 'we all know we need a balanced diet during pregnancy, but your body will dictate what you will eat'.

Food practices during breastfeeding

Breastfeeding diet advice was drawn mainly from the lifeworld as noted earlier. This advice applied to the entire duration of breastfeeding. The shortest duration of breastfeeding was two months; the longest eighteen months. While some women reported following this long-term cultural diet prescription a number of women found it monotonous and restricting. Soups, especially palm-nut soup, were singled out for attention for these reasons and also because many women found them excessively fattening.

> I almost ate everything with soup; groundnut soup, palm nut soup. Essentially every meal was taken with soup, rice balls, TZ, rice and soup. I was getting tired of it. (Dagao woman, aged thirty-two, banker, one child)

> You were advised to eat palm-nut soup every day. Can you imagine eating palm nut soup every day for a year? You just put on weight. When they [female relatives] were around you found ways of reducing the amount you ate. After they left you would stop eating it daily. (Asante woman, aged thirty-six, pharmacist, one child)

The breastfeeding period was a significant period for intergenerational bonding and tensions. In Ghana there is a strong tradition of elderly women managing processes of childbearing (Fayorsey 1992/1993; Guerts 1997). One Asante woman noted that mothers would often say to their daughters '*didi na wo*' (eat so you can give birth) prior to and during first pregnancies. This notion of '*didi na wo*' was articulated in various forms to other young women (in their twenties and thirties) in this study. Mothers and/or aunts do not only provide nutrition advice during pregnancy but also take care of their daughters and/or nieces after childbirth in their own homes or in their daughters'/nieces' homes for up to one year. These elderly female relatives cook, clean and look after the newborn, giving the new mother time to rest and regain pre-pregnancy strength and well-being. This period is therefore more strongly structured by cultural norms and practices regarding childbearing compared to the often isolated period of pregnancy when a woman's experience is circumscribed predominantly by nuclear family relations.

This period also provides insights into the way women respond to cultural norms and practices regarding childbearing. With the exception of one woman who had all her children in Canada,[9] all the women had their mothers or aunts take care of them and their babies. The duration of assisted care ranged from two weeks to one year. Assisted care was usually longer for first births. Mothers performed the full range of duties outlined earlier. This period therefore represented a significant phase of decreased physical activity for many women. For others assisted care freed them to return to work soon after childbirth.

> Our mothers and aunts would take our babies and say *ka da* (go to sleep, or go get some rest). They would stay for months and you could go back to work soon after your child was born. (Fante woman, aged sixty-four, retired, three children)

> My mother spent a month with me. She did the washing and cooking so I had some resting time. (Dagao woman, aged twenty-nine, NGO administrator, one child)

Many women attributed excessive weight gain during this period to regular consumption of fattening foods and decreased physical activity.

4. Weight management during pregnancy and breastfeeding

All the women believed childbearing signified a period of weight gain. During pregnancy many women increased food portions and frequency of eating largely due to cravings. With the exception of two women, all women reported weight gain during pregnancy. Not all the women who gained weight remembered the exact figures.[10] A young northern woman gave a typical response by this group: 'Yes, I gained weight but could not say by how much.' Two described their weight gain in terms of increased dress sizes. Of the ten women who remembered exact weight changes, weight gain ranged from 2 kg to 15 kg.

The breastfeeding period posed similar weight gain risks. Four factors were cited. Three centred on complex dynamics of food consumption; the final related to physical activity. First, for many women the dominant foods recommended for breastfeeding were highly fattening foods. Eating fatty soups every day for instance contributed to weight gain. Second, some women experienced increased appetite during breastfeeding – 'breastfeeding makes you hungry' (one woman noted) – and maintained high levels of general food consumption during this period. Third, the demands of childcare – especially the frequency of breastfeeding and tending to crying babies at irregular periods – changed food consumption patterns:

> We gain weight because we do not eat at scheduled times. You breastfeed late, you are very hungry, you eat and then go to bed. You might have to take care of babies at night or very early in the morning. All these contribute to irregular eating habits. (Ewe woman, aged fifty, University lecturer, three children).

The final reason was a general decrease in physical activity due to assisted childcare provided by mothers and older female relatives. For two women, complications following a caesarean section prevented them from engaging in strenuous physical activity after childbirth. For these complex reasons some women felt the breastfeeding period presented greater risks for (excessive) weight gain compared to the pregnancy period.

It did not appear that women made a conscious effort to manage weight during pregnancy. The prevailing attitude was to be well nourished during pregnancy in order to have a healthy baby. Consciousness about weight management during pregnancy was generally mediated by health professionals. All the women who had to manage their weight during pregnancy did so after being told by their doctors. There was greater consciousness to manage weight during breastfeeding for some women. This was influenced largely by the recognition that breastfeeding foods were fattening. These women adopted common sense strategies to reduce or cease consumption of fattening breastfeeding foods. A few women increased physical exercise, mainly by carrying out more household chores perceived as physically exhausting (cleaning and washing). These largely self-driven weight management strategies were clearly opposed to the cultural prescription for more milk producing foods and for rest. Expert scientific advice on health and weight management was missing during this period. Several women had been unable to return to their pre-childbearing weight. Two women who had had pregnancy-induced hypertension had developed adult onset hypertension, which persisted at the time of interviews.

Conclusion

This study aimed to explore the extent to which the childbearing life-stage presents obesity risk for Ghanaian women and what this means for women's diet, weight and health management strategies during and after childbearing. Three questions were posed:

(1) How does culture structure experiences of childbearing?
(2) How do women respond to these cultural factors during the childbearing period and what mediates these responses?
(3) What are the implications of (1) and (2) for diet, weight and health management interventions for women?

I focus on the key insights relating to the research questions and consider the place of this study within the broader literatures on obesity and chronic disease.

How does culture structure experiences of childbearing and how do women respond to these cultural factors?

From the women's own accounts the childbearing period constituted a period during which cultural pressure (from mothers, older female relatives and acquaintances) to gain weight and sustain weight gain increased. However, the research focus on cognitive polyphasic processes yielded four key insights about the complex relationship between cultural pressures and women's everyday experiences. First, cultural knowledge coexisted with scientific and body-self knowledge as the dominant modalities women drew on during the childbearing period. Second, the availability of information from cultural and scientific sources varied across the two periods of childbearing. The pregnancy period was characterised by a more balanced co-existence of cultural, scientific and for some women, scientised knowledge (see first section of results). Women drew eclectically from these sources to make sense of and manage pregnancy experiences.

The breastfeeding period was characterised by the dominance of cultural knowledge, through the physical presence of mothers and elderly female relatives in the lives of women, and the corresponding absence of health expert communication. Thus culture exerted its influence more strongly during breastfeeding than during pregnancy. Third, cultural knowledge was not entirely reified: for many women cultural knowledge constituted one knowledge resource in a range of other knowledge resources; for some women aspects of cultural knowledge were deemed illegitimate and potentially detrimental to maternal health. Finally, body-self knowledge emerged as an important mediator in everyday experiences of pregnancy and breastfeeding. During pregnancy the unpredictability of physiological processes overrode external information (from cultural and scientific sources) on diet and health practices. During breastfeeding recognition of the impact of unhealthy eating practices influenced strategies of resistance to cultural prescriptions on diet and health practices.

This study presents findings that are aligned with current discussions on the impact of culture on female obesity in Africa (Maletnlema 2002; Prentice 2006). However, like recent studies in Ghana (Duda et al. 2006), they draw attention to the relative power of culture and the presence of individual resistance and point to the need for nuanced analysis on the ways in which women respond to obesogenic cultures and environments in their everyday life and at specific life-stages.

What are the implications of these results for diet weight and health management interventions for women?

The childbearing period constituted a period when many women gained weight and when some developed weight-related pregnancy complications. For some women the breastfeeding period posed greater risks for weight gain due to an increase in consumption of fattening breastmilk-producing foods and a decrease in physical activity. Four women who had pregnancy-induced hypertension currently live with hypertension. While their pregnancy-induced hypertension was not attributed to weight gain, the women implicated their inability to manage their diet and weight during breastfeeding and after childbearing on their current health status.

The study demonstrates that there are different challenges for the pregnancy and breastfeeding phases. For many women, the pregnant body constituted the ultimate guide to what one thought, felt and did about food daily during pregnancy. This body-self knowledge over-rode other legitimate sources of knowledge on – the right or wrong – pregnancy foods, including fattening foods. From the positive end, physiological changes in the pregnant body enhanced appetite for simple healthily cooked foods. From the negative end, extreme physiological changes caused either under-consumption or over-consumption of nutritious or non-nutritious food. This common experience has two broad implications. First, it implies that embodied knowledge and its everyday functions have to be prioritised and legitimised in ante-natal care. Second, it suggests that pregnant women whose experiences lie at opposite ends of the physiological disruption continuum (e.g., loss of appetite versus increased appetite) are more likely to be non-compliant with diet weight and health management interventions and therefore require greater individualised medical/nutrition support.

The greatest challenge facing women during breastfeeding was the cultural prescription to eat fattening foods on a daily basis for the duration of breastfeeding and reduce physical activity for a number of months after childbirth. Many women found the breastfeeding diet monotonous, restricting and fattening. Some employed resistance strategies – reducing food portions, ceasing daily consumption of fattening soups, and taking up household chores – to minimise the impact of the prescribed diet on their body size. However these strategies were fundamentally self-derived. They operated within a vacuum of expert scientific advice on diet weight and health management.

Yet, it is this during this period, when some women resisted cultural prescriptions of diet and health management, consciously sought to manage their weight and adopted common sense strategies to achieve this goal, that expert advice was most needed. There is a clear need for post-natal interventions on food practices and physical activity that: (1) help women reduce weight gained during pregnancy and breastfeeding; (2) offer a practical guide for diet weight and health management for women after childbearing; and (3) actively involve influential older female relatives. The evidence that Ghanaian women privilege expert maternal health knowledge generally and resist problematic cultural prescriptions for postnatal care suggests that postnatal interventions targeting women's diet, weight and health management stand a chance of success.

Acknowledgments

I would like to thank Macarius Donneyong and Kojo Ayernor for providing research assistance. The pilot phase for this study was presented at the Oxford Institute of Social and Cultural Anthropology (ISCA) Workshop on 'Fatness, Food and Childbearing: Cultural Perspectives on the Body, Nutrition and Reproductive Practices', in October 2006. I am grateful to the workshop participants whose comments and critiques informed a revised study methodology.

Notes

1. There is evidence of some urban Ewe women returning to their rural hometowns to deliver their children (D. Badasu, personal communication, May 2007). In this study one northern woman returned to her hometown to receive childcare support from her mother as well as post-natal services.

2. Several studies suggest Ghanaian women, both rural and urban, do multiple work (cf. Allman and Tashjian 2000; Avotri and Walters 1999; Clarke 1994; Fayorsey 1992/1993). It was important therefore to pay attention to the nature and range of participants' work.

3. The 'stopping criterion' for qualitative research: the point at which no additional variety of meaning is gained from data collection. As Gaskell (2000: 43) observes:

   The first few [interviews] are full of surprises. The differences between the accounts are striking and one sometimes wonders if there are any similarities. However common themes begin to appear and progressively one feels increased confidence in the emerging understanding of the phenomenon. At some point a researcher realizes that no new surprises are forthcoming.

   The point at which no new surprises emerge is the point of meaning saturation.

4. Note-taking was preferred because it created a more relaxed interviewing setting, especially for opportunistic interviews. For some interviews note-taking was the only practical option, for example with market women interviewed during their working hours.

5. Following the original development of the concept of lifeworld by Husserl (1970), lifeworld is used here to mean 'the lived world'. This encompasses the embodied intrasubjective experience, intersubjective relations and the material and symbolic dimensions of everyday social life that impinge on intra and intersubjectivity.

6. The women who mentioned Milo stated that this was a specific diet prescription not interchangeable with other beverages such as tea.

7. Pica has been defined as 'a perverted appetite for substances not fit as food or of no nutritional value' (*Stedman's Medical Dictionary* twenty-sixth edn., cited in Goldstein 1998: 465).

8. Badasu (2004: 25) outlines a range of prohibited foods during pregnancy among the Ewe: okro and ripe plantain because they are 'slippery' and may cause miscarriage; snails because they make a child salivate too much and crabs because 'they are believed to make children walk in a clumsy manner'. This suggests that taboos around some pregnancy foods – in this case snails and okro – are shared across ethnic groups.

9. The traditional system of assisted childcare does travel abroad, particularly for Ghanaian women with financial security or from

wealthy families. It is well known within local and international Ghanaian communities that mothers often travel to Europe and the U.S.A. to provide assisted care to their daughters after childbirth. In the case of this Canadian-based woman, assisted care was provided by older female members of the close-knit diasporan Ghanaian community to which she belonged during her stay in Canada.

10. It is difficult to know whether this lack of knowledge is a function of the study methodology or to emerging evidence that knowledge of actual weight is poor in Ghanaian communities. A psychological study that examined cultural perceptions of body size and weight management strategies among one hundred Accra-based men and women found that the majority did not know their weight or height (Anum and de-Graft Aikins, forthcoming).

References

Acquah, I. 1958. *Accra Survey*. London: University of London Press.

Agyei-Mensah, S. and A. de-Graft Aikins. 2007. 'Epidemiological transition and the double burden of disease in Accra'. 2007. Paper submitted to Union for African Population Studies (UAPS) Conference, 2007; available at: http://uaps2007.princeton.edu/abstractViewer.aspx?submissionId=70383.

Allman, J. and Tashjian, V. 2000. *'I Will Not Eat Stone.' A Women's History of Colonial Asante*. Oxford: James Currey.

Amoah, A.G.B. 2003. 'Sociodemographic Variations in Obesity among Ghanaian Adults', *Public Health Nutrition* 6(8): 751–75.

Angel, R. and Guarnaccia, P. 1989. 'Mind, Body and Culture: Somatization among Hispanics', *Social Science and Medicine* 28: 1229–38.

Anum, A. and de-Graft Aikins, A. Forthcoming. 'Cultural Perceptions of Body Size and Weight Management Strategies among Urban Ghanaians: Implications for Dietary and Nutrition Interventions'.

Avotri, J.Y. and V. Walters. 1999. ' "You Just Look at Our Work and See If You Have Any Freedom on Earth": Ghanaian Women's Accounts of Their Work and Their Health', *Social Science and Medicine* 48(9): 1123–33.

Bartlett, F.C. 1932. *Remembering: A Study in Experimental and Social Psychology*. Cambridge: Cambridge University Press.

Bates, M.S., L. Rankin-Hill and M. Sanchez-Ayendez. 1997. 'The Effects of the Cultural Context of Healthcare on Treatment of and Response to Chronic Pain and Illness', *Social Science and Medicine* 45(9): 1433–47.

Biritwum, R.B., J. Gyapong and G. Mensah. 2005. 'The Epidemiology of Obesity in Ghana', *Ghana Medical Journal* 39(3): 82–85.

British Medical Journal. (2005). 'Health in Africa', 331: 7519.

Busia, K.A. 1950. *Report on a Social Survey of Sekondi-Takoradi*. London: Crown Agents.

Clark, G. 1994. *Onions Are my Husband: Survival and Accumulation by West African Market Women*. Chicago/London: University of Chicago Press.

Dake, F.A. 2009. 'Socio-demographic Correlates of Obesity among Ghanaian Women'. Paper presented at the Thirtieth African Health Sciences Congress, Accra, Ghana. 15 September.

de-Graft Aikins, A. 2004. 'Social Representations of Diabetes in Ghana: Reconstructing Self, Society and Culture', Ph.D. dissertation. University of London: London School of Economics and Political Science.

———— 2005. 'Healer-shopping in Africa: New Evidence from a Rural-urban Qualitative Study of Ghanaian Diabetes Experiences', *British Medical Journal* 331: 737.

———— 2007. 'Ghana's Neglected Chronic Disease Epidemic: A Developmental Challenge', *Ghana Medical Journal* 14(4): 154–9.

Dixit, A. and J.C. Girling 2008. 'Obesity and Pregnancy', *Journal of Obstetrics and Gynaecology* 28(1): 14–23.

Fayorsey, C.F. 1992/1993. 'Commoditization of Childbirth: Female Strategies towards Autonomy among the Ga of Southern Ghana', *Cambridge Anthropology* 16: 3, 19–45.

Flick, U. 1998. 'EverydayEvery day Knowledge in Social Psychology', in U. Flick (ed.), *The Psychology of the Social*. Cambridge: Cambridge University Press, pp. 41–59.

Gaskell, G. 2000. 'Individual and Group Interviewing', in M. Bauer and G. Gaskell (eds), *Qualitative Researching with Text, Image and Sound: A Practical Handbook for Social Research*. London: Sage, pp. 38–56.

Gervais, M., N. Morant and G. Penn. 1999. 'Making Sense of "Absence": Towards a Typology of Absence in Social Representations Theory and Research', *Journal for the Theory of Social Behaviour* 29(4): 419–44.

Ghana Statistical Service (GSS), Noguchi Memorial Institute for Medical Research (NMIMR), and ORC Macro. 2004. *Ghana Demographic and Health Survey 2003*. Calverton, MD: GSS, NMIMR, and ORC Macro.

Goldstein, M. 1998. 'Adult Pica: A Clinical Nexus of Physiology and Psychodynamics', *Psychosomatics* 39(5): 465–69.

Guerts, K.L. 1997. 'Well-being and Birth in Rural Ghana: Local Realities and Global Mandates'. Paper presented at the Fifth Annual Penn African Studies Workshop, 17 October. Retrieved 4 May 2007 from: http://www.africa.upenn.edu/Workshop/geurts.html.

Helman, C. 2000. *Culture, Health and Illness*. Oxford/Boston: Butterworth-Heinemann.

Hill, A.G., et al. 2005. 'Self-reported and Independently Assessed Measures of Population Health Compared: Results from the Accra Women's Health Survey'. Paper presented at the International Union for the Scientific Study of Population (IUSSP) General Conference, France, 21 July.

Husserl, E. 1970. *The Crisis of the European Sciences and Transcendental Phenomenology*. Evanston, IL: Northwestern University Press.

Kruger H.S., et al. 2005. 'Obesity in South Africa: Challenges for Government and Health Professionals', *Public Health Nutrition* 8: 491–500.

Kumi-Aboagye, P. 2008. 'Status of MDG 5 – Evidence from the Field'. Paper presented at the National Consultative Meeting on the Reduction of Maternal Mortality in Ghana, Accra, 8 July.

Levine, C.E., et al. 1999. 'Working Women in an Urban Setting: Traders, Vendors and Food Security in Accra', *World Development*, 27(11): 1977–91.

Lupton, D. 1998. *The Emotional Self: A Socio-culturalSociocultural Exploration*. London: Sage.

Maletnlema, T.N. 2002. 'A Tanzanian Perspective on the Nutrition Transition and its Implications for Health', *Public Health Nutrition* 5(1A): 163–8.

Messer, E. 1989. 'Methods for Studying Determinants of Food Intake', in G.H. Pelto, P.J. Pelto and E. Messer (eds), *Research Methods in Nutritional Anthropology*. Tokyo: United Nations University.

Ministry of Health (MOH) (Ghana). 2001. *The Health of the Nation: Reflections on the First Five Year Health Sector Programme of Work, 1997–2001*. Accra: MOH.

Moscovici, S. 1961/1976. *La Psychanalyse: Son image et son public* (second edn. 1976). Paris: Presses Universitaires de France.

———— and G. Duveen 2000. *Social Representations: Explorations in Social Psychology*. New York: New York University Press.

Popkin, B.M. 1998. 'The Nutrition Transition and its Health Implications in Lower-income Countries', *Public Health Nutrition* 1: 5–21.

———— 2005. 'Using Research on the Obesity Pandemic as a Guide to a Unified Vision of Nutrition', *Public Health Nutrition* 8: 724–29.

Prentice, A.M. 2006. 'The Emerging Epidemic of Obesity in Developing Countries', *International Journal of Epidemiology* 35: 93–99.

Rose, D., et al. 1995. 'Questioning Consensus in Social Representations Theory', *Papers on Social Representations* 4(2): 1–155.

Siervo, M., et al. 2005. A Pilot Study on Body Image, Attractiveness and Body Size in Gambians Living in an Urban Community. *Body Image*.

World Health Organization (WHO). 2005. *Preventing Chronic Disease. A Vital Iinvestment*. Geneva: World Health Organization.

World Health Organization/Food and Agriculture Organization (WHO/FAO). 2003. *Diet, Nutrition and the Prevention of Chronic Diseases: Report of a Joint WHO/FAO Expert Consultation*. Geneva: World Health Organization.

Chapter 8

UNHEALTHY, UNWEALTHY, UNWISE:
SOCIAL POLICY AND NUTRITIONAL EDUCATION IN A DISADVANTAGED COMMUNITY IN IRELAND

Shauna Clarke

Introduction

Obesity has come into the foreground of political consciousness and has invoked reactions in many forms. In February 2007, it has been brought to our attention in the form of the widespread media coverage of an English boy who weighed 218 pounds: said to be three times the weight of a healthy child his age. Authorities in London considered taking the obese eight-year-old into protective custody due to neglect. The conflict revolved around issues of the assertion of power and the use of certain valued forms of knowledge. In short, authorities believed that they had the power to separate a child from his mother because she was seen as not fulfilling her *'proper role'* by providing and caring for her son, because of a lack of knowledge regarding his food and nutritional intake. In Ireland efforts have been made to tackle cases like this. A general concern over fat and its negative impacts has resulted in the creation of The National Taskforce on Obesity (NTFO). With rising obesity rates, Ireland is beginning to take stock of its situation and to see tendencies towards being overweight as a threat that needs to be addressed.

This paper explores the structure of Irish government policy on obesity and its influence on nutritional health promotion. I will discuss this with reference to my own research which is based in North Clondalkin, a disadvantaged area in south-west Dublin. This research focused on a community nutrition project called 'Healthy Food Made Easy' (HFME) run by dieticians and trained 'peer leaders' drawn from the local community. This research was mainly a product of participant observation, during the classes, and meetings with individuals outside of the classes.

The contact and discussion with inhabitants of, what had been termed, a disadvantaged community acts to reveal the voices and experiences of these women. My mode of research allowed me to interact with them in their daily lives and provided a forum for their concern over eating habits and its effect on health. They provided a focus and illustration of the real people living within the policy initiatives that I was investigating, revealing the struggles of balancing everyday life with the wealth of information on healthy eating.

Through the women in the HFME groups, I will examine the assimilation and interpretation of nutritional education and its influence on the participants' everyday food practices and the production of discourse concerning food and the body. This paper looks at the power relationships between the state and the individual around the policy of healthy eating, especially the individualising quality of the state discourse on the subject and rejection of other forms of knowledge production about health and food.

Obesity is as much a social phenomenon as a nutritional one. For the first time in history, plumpness and obesity are signs of poverty. The wealthy can afford to eat right, exercise right and watch their weight. They possess the cultural capital (Bourdieu 1986) of knowing how to manage their fat: however disadvantaged communities are perceived as uninformed and lacking in cultural capital. The disadvantaged do not belong to the correct habitus (Bourdieu 1977) and therefore expose themselves to technocratic management and interference from institutional bodies. Crotty (in Coveney 1998: 426) sees the control of the discourse on health and eating by nutritionists as a form of social control, which attempts to ensure that people follow the rules that are seen as acceptable. Crotty sees nutrition's 'scientistic' control as based on a population that is assumed to be 'sick' and in need of reform. Through this lens I will look at the HFME course as a response to fears of an obesity epidemic in Ireland and the education of those most at risk of becoming overweight: the lower classes living in a disadvantaged area.

An 'obesogenic' environment

In order to situate my discussion of policy initiatives and the intervention of health education I feel it is necessary to introduce the physical and resulting social environment in which these were taking place. The subject area, of North Clondalkin, that I located my research in is a perfect example of a susceptible infrastructure, social structure and resulting ideology. The planning of North Clondalkin was based on the philosophy of the 'neighbourhood unit' and it was envisioned that 'small scale arrangements of dwellings could permanently enrich community life by stimulating frequent interactions between residents and engendering a spirit of neighbourliness through spontaneous co-operation'. (Bartley 1999b: 228). The planning aimed for the centralisation of community resources with schools and commerce centres within a five-minute walk of their local catchment area. However, this was not the outcome of the spatial distribution. In the planning of Clondalkin, the Rowlagh/Neilstown area, was marked as a strategic development zone, a planning objective by the Government that attempted to solve the social housing problem but is now used as an example of a prototype gone wrong, resulting in ghettoisation.

My first experience of fieldwork showed me the isolating effect of this landscape firsthand. On a dark Tuesday night I had an impromptu tour of the Rowlagh/Neilstown area of North Clondalkin. I was on my way to my first HFME class in which I would be doing participant observation on community nutritional education. The class was run in a community-owned house within a large housing estate. Upon entering the estate I followed the directions I had been given, which seemed simple enough, however my car came to a dead end before the directions had finished. Unfazed, I retraced my steps and attempted a slightly altered route, but there were no houses like the one that I was looking for. After asking several people for directions, and many three-point turns later, I finally arrived at our destination a half hour late. Here I came face to face with the seclusion that often accompanies the planned structure of the estates: maze-like roads edged by monotonous, similar looking houses. Roads leading nowhere, taking you farther away from the main road and access to the outside.

This experience helped me to understand the isolation and separation of these estates and the people who live in them, not just from the outside, but from each other as members of a disjointed community. As Brendan Bartley comments:

> The area has the restricted inward focus of an isolated enclave. This
> implosive quality is reinforced by the fragmentation effect created by
> the internal design of the neighbourhood housing estates with their
> emphasis on separateness: separate neighbourhoods, separate estates,
> separate house clusters and cul-de-sacs. (Bartley 1999b: 242)

Bartley (1999b: 235) describes the area as littered with run-down
neighbourhood centres, public buildings invariably surrounded by
large palisade fencing and shuttering, housing estates that face
inward, turn their back on public areas and are devoid of facilities.

This first experience of fieldwork in North Clondalkin is an
introduction to not only the physical environment but the resulting
social context from which the people with whom I worked came
from. I think that the many cul-de-sacs that I encountered can be
compared to the many dead-ends that they often face. North
Clondalkin, but especially Rowlagh, is considered a disadvantaged
area. This can be seen in the lack of facilities present in the
surrounding area. In relation to accessing adequate nutrition, one of
the major concerns of this paper, the structure of the transport in
North Clondalkin leaves residents a long way from appropriate
sources, the nearest supermarket being three miles away with no
direct bus route.

Friel and Conlon (2004b: 25) discuss the direct affect of the
physical environment on dietary behaviour. 'The type of shops that
people can reach is based on their physical capabilities and transport
options and access to such facilities may be restricted if car use is not
an option and adequate public transport is not provided.' In an area
designed intentionally to cater for the needs of the private car,
vehicular movement still remains congested. The estates with the
most road space, the local authority estates, are least able to take
advantage of the potential of the wider roads systems as these areas
have the lowest level of car ownership. All these aspects contribute
to the social deprivation of the area and the continuing isolation of
the communities. Clondalkin is desperately in need of more facilities,
which are needed to fill the gaps in the social, recreational and
cultural life.

Everyday situations pose a difficulty to the residents of North
Clondalkin. A task as simple as doing the weekly food shopping is
complicated through lack of adequate facilities. Access to large
supermarkets with the greatest choice of foods at lower prices
becomes difficult if a subject lives in the middle of one of the large
housing estates in Rowlagh. Without a car, simply getting from your

door to a bus stop is not an easy proposition. The long way involves walking through a maze of houses, walking twice the length of the journey due to the fact you must follow the road around the houses. The short way is an across-country trek, over the hilly green towards the road. This may seem like the easy option but in the winter it is a hike through slippery mud while laden down with heavy bags of shopping, perhaps with a young child in tow. Once out on the road you must still make the journey to the supermarket via bus and back with your load of shopping. This will surely take up a considerable amount of your day. Another option is the local shop, although limited in its choices: at least this venture can be achieved without spending the majority of the day undertaking the task and if all the goods cannot be carried home in one trip then it is close enough to go back for the rest later. These structures of the environment act to restrict people's choices and movements in normal everyday situations and to impose further difficulties that limit the response to these boundaries.

Situations like the ones I have described act to frame my discussion in terms of a disadvantaged area and community. Through this context I want to investigate the negotiation of power relations through the medium of nutritional education. The term 'food desert' has been used to describe areas of relative isolation (Reisig and Hobbiss 2000 in Friel and Conlon 2004b: 25). The physical and economic barriers experienced in North Clondalkin contribute to a 'food desert' by the lack of access to healthy food. The environment in which the disadvantaged subjects lives and its intersection with policy initiatives is a helpful way to look at the societal discourse and authoritative knowledge of obesity and fat in Ireland.

Health education, policy and the individual

The HFME course is in part a response to fears of increases in obesity levels in Ireland. The 2005 report by the National Obesity Task Force discusses the determinants of obesity in Irish society. Factors include: a decline in demanding physical work, sedentary lifestyles and reduced leisure-time activity, increase in commuting times to and from work and the widespread use of motor travel rather than active transport and a change in eating habits characterised by an increase in convenience foods and a decrease in home cooking.

Despite all these influencing external factors on an individual's lifestyle, and the impact that these have obviously had, themes of choice, rational action (in regard to health) and individuality dominate the report prepared by the NTFO. The effect of the 'obesogenic' environment (obesity-promoting environment) (NTFO 2005: 70) is acknowledged in its possible role and influence on obesity levels: however, suggestions for change and improvement tend to centre on the individual.

The NTFO states that 'At the centre of these changing environments is the adult or child who requires certain skills to enable him or her to make "healthy" choices for life' (NTFO 2005: 70). It is clear in the report that the emphasis and prompting of change is placed onto the individual living within this environment and not on significant change in the structure of a disadvantageous environment.

The 2000 to 2005 National Health Promotion Strategy discusses the major determinants of health. It recognises the role of the individual and environment.

> Many factors influence and determine health, whether at an individual or population level [as seen in Figure 8.1]. Social, economic and environmental factors are the main external or structural determinants of health. At an individual level, factors such as age, sex, hereditary factors and lifestyle choices are important. (Department of Health and Children 2000: 10)

This may seem like an equalising of the responsibility of both the individual and environment. However, by taking a closer look at Figure 8.1 that illustrates this information, one can see a hierarchy of responsibility. This diagram presents the information in a way as

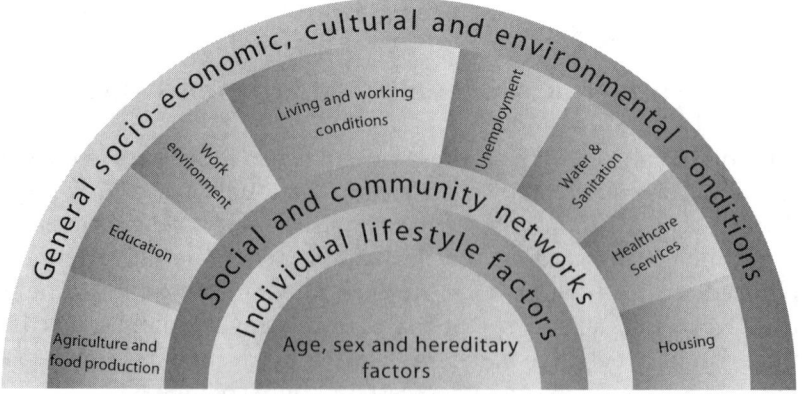

Figure 8.1: *The major determinants of health*
Source: Department of Health and Children 2000: 10.

to create the individual as the core and to work outwards towards the environment, rather than the environment being the basis or being on a horizontal level of impact with the individual. In the same way that Jordan (Davis-Floyd et al 1997: 17) argues for a move from a situation where authoritative knowledge is hierarchically distributed to a situation where it is, by consensus, horizontally distributed – where all participants contribute to the store of knowledge on the basis of which decisions are made – so too should the degree of accountability for the individual be, at least, equalised with the environment.

This diagram acts to identify individuals as the main factor in determinants of health and therefore the core problem and greatest responsibility to negative health. The individual is either facilitated or limited by their surrounding environment. However the obesogenic environ of Ireland seems to have a negative impact on all, even if the NTFO do not attribute proportionate blame to it. In the case of the lower classes, environment further problematises the individual. The reality is that the achievement of physical and mental well-being is not the responsibility of the individual alone. People's ability to pursue good health is limited by varying degrees of skills, information and economic means. [Department of Health and Children 2000: 10]

The significance placed on the 'problem individual' within health promotion is to be addressed through education in the necessary areas. In the case of obesity the focus is on diet and physical activity. In the previous diagram, although education is placed within the realm of environment and is deployed within a specific context, it operates on an individual level through health interventions and community education models. In Foucault's terminology this is called the 'anatomo-politics of the human body' (Foucault 1990: 139) where methods of controls and interventions are exerted on the individual by larger, powerful institutions. 'A Health Promotion Strategy: making the healthier choice the easier choice', produced by the Department of Health in 1995, offers the following statement of ideal health promotion:

> Health promotion at an individual level involves educational processes enabling people to acquire information and skills that will help them in making good decisions in relation to their health. At a community, regional and national level it involves the development of appropriate policies, structures and support systems so that the healthier choice becomes the easier one to make. (Department of Health 1995: 2)

This concentration on the individual level has other consequences. It pictures society as populated by rational decision-making individuals, who, regardless of the social and cultural environment, wish to prioritise health over other considerations. This subject is an autonomous rational ego who uses expert systems (nutritional science) reflexively to regulate everyday life (Petersen 1997: 190). The relation of these individuals to society, as a whole, and to the environment in which they are living is not always clear. Individualisation problematises the individual, constructing the individual as the locus of the problem of ill health and disease, and offering individual behaviour change as the only way of dealing with the problem. The phenomenon of 'healthism' (Greco 1993: 357 cited in Petersen 1997: 198) regards the individual as having a choice in preserving him or herself from disease and ensuring their continued health by regulating risky behaviour through their use of knowledge.

Knowledge is produced and understood in many different ways in different social settings. A 'politics of knowledge' is present in every situation where information is disseminated. This 'politics of knowledge' is a hierarchy of knowledge that elevates some forms of knowing, authoritative knowledge, over others, subservient knowledge. Both individuals and groups participate in the representation of information and the framing of reality through techniques of representation. Claims to true forms of knowledge, or 'facts', are justified through certain forms of validation and influenced through prioritised structures and beliefs within culture.

The interaction between power and knowledge can be seen in its practical application through the subjects' experiences in the HFME nutritional education course. This negotiation allows me to examine the intervention of health promotion on knowledge and expressions of knowledge of the involved community and the relationship between both 'expert' and 'lay' knowledge. One of my respondents, Brenda, is a perfect example of how the authority over the ownership of knowledge between subjects and peer leaders was challenged. Brenda, from the Literacy Training Initiative (LTI) class, is an outspoken woman in her early forties who has a large family. She holds certain beliefs regarding food and eating, which present a challenge to the notion of health, as defined by medicine and the state via health promotion. Brenda defines healthy eating as listening to what her body tells her to eat. She believes that she is a healthy eater because, 'I eat when my body tells me to eat. When I need something, some nutrient or something, I'll crave it. My body knows what type of food it needs and it tells me what I need to eat.'

Brenda feels that through her experience of being a mother she has accumulated a great deal of knowledge and feels that the HFME course is only useful in order to provide a better understanding of food. It is pointless in its imposition of informing subjects on what and what not to eat, as she relies on her body for this information. Brenda believes that her body contains all the knowledge that she requires and that it has ways of telling her its needs, for example through food cravings or bad reactions to certain types of food.

This idea of the body as a source of authoritative knowledge contests the authoritative knowledge of rational health choices based on science. Brenda's belief in her bodily knowledge contests the authority of the scientific discourse of nutritional science. According to a biomedical model of the understanding of the body, Brenda's account is seen as having no rational basis and is instead based on the whims of the body. Within the structure of the HFME course this type of experiential knowledge was readily accepted if it complied with the information and aims of the course, even if it did not come from an accepted source, the body. This lived experiential knowledge was viewed as unreliable compared to rational scientific knowledge of food and the processes of the body, as defined by medicine. It needed to be informed and supported by scientific fact and logically analysed in relation to this in order for it to be judged as trustworthy.

In this case, health education in its aim to produce a body that is docile to the logical mind has failed, owing to the conquest of the rational by Brenda's bodily experience. Brenda's declaration of her embodied bodily knowledge as superior, to any learnt or scientific form of knowledge, instils it with power and authority in response to the authoritative knowledge of nutritional science. It defies Gastaldo's notion (1997: 118) that 'Radical and traditional health education share an underlying notion of empowerment through education or subjection through ignorance.' Brenda subverts this and uses her body and its embodied knowledge as an instrument of power. Through this she reclaims her body and empowers it with 'real' knowledge, that which she experiences for herself, and rejects the overriding structure of medical knowledge. She uses this as a way of reclaiming her body from authority and claiming it and its production of knowledge as her own and as of the highest order. This contests the notion of her construction as an un-knowledgeable and powerless subject because she constructs herself as a rational actor by using the discourse of health. She manipulates this discourse of health to imbue her body with powerful knowledge that does not need the dominant knowledge of medicine.

In the discussion of the competing discourses of health I do not wish to succumb to an argument in the form of the medicalisation critique or to indulge in a Foucauldian interpretation of the docile body where discourses subjugate human agency. Instead I wish to emphasise the 'lived experience' of the human body and the negotiations between experience and discourses of power and authority in the form of health promotion. It is important to look at ways in which members of the lay population respond to authoritative medical discourse and utilise and negotiate it in their everyday lives. These strategies of power become local techniques and are exercised at the level of everyday life.

Brenda's discourse of resistance is an excellent example of this. It is a response to a number of mitigating factors that influence the way in which authoritative knowledge is subjugated and devalued in order to assert a valuable subject position in the face of authority and ownership of knowledge.

'Empowered' with knowledge, food habits and consumption becomes a matter of choice: the choice to act rationally, utilise acquired knowledge and consume responsibly or to indulge in deviant 'irrational' behaviour. Individuals are empowered with choice and are therefore to be held responsible if they choose wrongly and suffer negative health consequences. Outside factors that might influence choice are flattened out of the representation. Health (the avoidance of disease) is seen as the only rational choice and therefore is the only choice once people have been informed of it. Just as authoritative knowledge is flattened and excludes other forms of competing knowledge that contends its ideas, so too is communication of this knowledge de-problematised. Knowledge produced by medicine is represented as being unbiased and value-free and so the presentation of this knowledge goes unquestioned and is to be unconditionally accepted as fact.

The choice of 'right' or 'wrong', 'good' or 'bad' foods is one that the subjects faced every day. Many of the subjects openly admitted to and broadcast their 'weakness' for certain foods: all of which were deemed unhealthy. Through the admittance of weakness or failure and the realization and understanding of these bad choices, through education, the subject takes control over their deviant practices and from now on can make informed and good choices. In this way they construct themselves through discourse, according to the technology of power, nutritional science. However one exception to this norm stood out in the HFME course. Johanna, who was in her late forties, was the eldest woman in the parenting group. She had a large family, who were grown up and had left home, except for her oldest son. She and her husband both shared cooking duties but she claimed that he did the majority of the cooking in the house as she was not interested in doing it now that the family had mostly left home. She was a thin woman who smoked a lot and claimed to have a small appetite. She said, 'Sure if he didn't cook, I probably wouldn't bother eatin' at all.' In the discussions and 'confessions' of bad foods Johanna disassociated herself from any foods that were deemed as unhealthy. She immediately removed guilt from herself and any bad lifestyle practices by confessing that 'I do not eat that anyway' to a variety of foods that were viewed in a negative light. While everyone else confessed to some weaknesses when her turn came, instead of the expected admission of guilt to some 'faux pas', she denied possessing any bad eating habits. Most of the subjects from the parenting course readily accepted the knowledge produced by biomedicine, the subjugation of their own knowledge and authority of other forms of knowledge. However, Johanna seemed to resist this by producing herself as already conforming to the standards and not in need of educating in these matters. Her continual denial of any bad food habits even led one of the other subjects to comment that 'sure Johanna doesn't need to be here at all, she knows everything and never does anything bad'. This construction of herself as already informed was a rejection of the external strategies of power that were placed on the subjects.

Other forms of rejection of the information evident in the classes were lack of interest, cooperation and participation in the group discussion. Each subject acted in a way to present themselves in a certain manner, as a certain 'type of person', engaged in 'rational' behaviour with a personal and emotional investment and reason behind their ideology and performance of self. Coveney states that 'The autonomy and choice of the "choosing" subject are always mortgaged to the moral principles that the subject has rationally set for itself" (1998: 462). This acts to illustrate the ways that hegemonic medical discourses and practices are taken up, negotiated or transformed by members of the lay population in their quest to maximise their health status and avoid physical distress and pain (Lupton 1997: 95). Johanna's expression of resistance defies the idea of the subject of authoritative knowledge as a helpless vessel for knowledge, but as a subject who may act and participate in their own construction of knowledge and self. This is also evident in the example of Brenda. Foucault's technologies of power and technologies of the self examine and theorise 'the ways in which people's experiences are controlled by others and the ways in which individuals control themselves' (Coveney 1998: 461).

Johanna used her claim to prior knowledge as an expression of her power within the class. Rather than challenging the ideas and power relations of the course she constructed herself within these ideals. Johanna presented herself as already embodied with the knowledge presented in class and already adhering to its principles. She did not need to be educated as she already practised the correct behaviours and assigned importance to the notion of health and correct eating.

Both examples of Brenda and Johanna produce themselves as active agents in control of their bodies rather than submissive subjects open to the imposition of authoritative knowledge and power. They reject the given power relations of the class and the subversion of the disadvantaged subject due to the assumption of their lack of knowledge. Johanna does this by using the ideology of health promotion and nutritional science to construct herself as a healthy individual already making healthy choices within the provided discourse. She sees herself as embodied with knowledge of health and therefore above and immune to the imposed authoritative knowledge of science and medicine. Brenda locates herself to the left of the provided discourse, valorising embodied knowledge and placing this above the authoritative knowledge of medicine, but constructs herself within the ideology of health, producing herself as a healthy body and individual.

Health promotion is a multi-level approach, requiring the interaction of individuals, communities, and regional and national institutions, yet someone has to make the decisions as to what is regarded as healthy. This is seen as down to the individual and where this fails, and the right or rational choice is not made, then it is seen as the responsibility of heath promotion to inform and educate these individuals or groups of individuals as to what the rational healthy choice is.

> The Taskforce's social change strategy is to give people meaningful choice. Choice, or the capacity to change (because the strategy is all about change), is facilitated through the development of personal skills and preferences, through supportive and participative environments at work, at school and in the local community, and through a dedicated and clearly communicated public health strategy. (NTFO 2005: 8)

According to Giddens (1991) in post-traditional society constraints over choice are effectively weakened, and the individual is confronted with a complex diversity of alternatives, especially in relation to 'lifestyle' (Petersen 1997: 191). Health promotion acts in reducing this unrestrained choice by providing a structure of 'good' and 'bad' or 'healthy' and 'unhealthy' choices. The individual is responsible for their choice. Through health promotion the individual is educated in what is the correct and rational choice, the 'healthy option'.

> Individuals must learn, 'on pain of permanent disadvantage', to conceive of themselves as the masters of their own fate, and to see events and conditions that happen to them to be a consequence of their own decisions. (Petersen 1997: 192)

The focus of social policy on the individual can be investigated through Brenda's reaction to this imposition of power and authority. Brenda succeeds in taking responsibility for her own actions while at the same time she rejects the authority of health promotion and positions herself and her body as the authoritative source of knowledge. She does this through her own discourses of rationality rather than utilising the rational provided by health promotion policy. The embodiment and possession of knowledge acts to advance her social power and status from docile, unknowing, disadvantaged subject to powerful actor in control of her own health. Brenda summed up her ideology on the issues of education on food and eating concisely. She said: 'it's good to know what you're eating, like what is in the foods and how this affects your body, but this is no good unless you actually listen to what your body is telling you is good or not'.

Negotiations of power

Throughout life different forms of knowledge compete for prominence as the authoritative voice. This was no different in my own experience of nutritional education. Authoritative knowledge is produced by the government and science in the form of health promotion, as a type of medical intervention. The contestation in my fieldwork formed mainly around issues of ownership of knowledge and the power that this imbued upon the possessor.

Within the HFME class, different forms of the negotiation of power and the acquisition of control were displayed through various manners of communication and structuring of discourse. Brenda's experience and reclamation of her own knowledge as powerful is one example of this. Another less subtle example was the questioning and rejection of the authority of the peer leaders which was displayed in the class.

Disadvantaged communities, like that of Rowlagh, experience the imposition of power through many forms of bureaucracy in every realm of life and oppose their disempowerment through various forms of resistance. The notion of nutritional education presupposes a lack of or undermining of the knowledge held by the subjects. This is especially relevant for the LTI groups as they do not choose to take the HFME course but it is a compulsory part of their programme. The re-conceptualisation of popular nutritional information into more correct and consistent forms by both peer leaders and dieticians is often interpreted as a subversion of the subjects' own knowledge and the imposition of the knowledge of nutritional science. The subjects feel disempowered and unsubstantiated in their position against the authoritative knowledge of nutritional science. In some ways they saw the HFME course as simply another form of technocratic management being imposed on themselves and their bodies. One of the subjects, Ruth, expressed her frustration at the intersection of all the different forms and sources of knowledge. 'Sure we don't really know anything. Ya can't eat this and ya can't eat that. All things are bad for you if you think about it anyway!'

Although the course aims to localise knowledge distribution through the use of peer leaders, within the model of community partnerships, it did not fully achieve its goal as this is still viewed as a form of power enforced from above. The peer leaders can no longer be peers due to their role of teaching, which requires the possession of dyadic knowledge. The peer leaders were also noticeably from a different background to the subjects. This removes them from the class, changing their status and instilling them with power. They may have come from the area of Clondalkin, but they were not from the community of Rowlagh and would not have been previously acquainted to any of the subjects. One thing that marked them as different was the variance between the accents of the peer leaders and those of the subjects. The classes reacted in different ways to this hierarchy of power. The LTI group who is obliged to take the course reacted to it by resisting the imposition of the power of the authoritative knowledge of nutritional science.

In the first LTI group the power struggle was particularly evident. The authority of the peer leaders was questioned and negated by the behaviour of a large part of the class. Most of the students in this class either ignored the peer leaders and what they were doing with the class or openly challenged their authority by using disruptive behaviour to interrupt and disorder the class. Jacinta, one of the younger members of the group, acted as main instigator of this resistance. Jacinta was a short, stocky, powerfully built woman, who wore tracksuits and had a macho attitude that was often displayed in class. Through vehicles such as doing other homework during the class, shouting across the room, refusing to answer any questions or voice opinions asked of her and suddenly leaving the classroom, Jacinta displayed her own power. By acting in this way she challenged the authority of the peer leaders and asserted her own authority on to the class. 'Health education is an experience of being governed from the outside and a request for self-discipline.' (Gastaldo 1997: 118). This was a response to the imposition of power from a top-down structure. The class followed in this behaviour as a way of contradicting their powerlessness as disadvantaged and un-knowledgeable subjects through the defiance of the authority of the peer leaders.

These subjects produced themselves in resistance to the authority of medicine, nutritional science and health promotion. It is through the politically charged ownership of power that they act and frame their agency within the class, in display to each other and the peer leaders. The behaviour of some of the subjects in the class was an open defiance of the imposition of authority and a rejection of the knowledge presented in the class.

Conclusion

Critser, at the end of his book, *Fatland*, comments 'How we get out of hell depends not on prayer, but rather upon a new sense of collective will: and individual willpower' (Critser 2003: 176). The hell that Critser refers to is that of rising levels of obesity. I find it hard to believe that Critser, after writing an articulate and well researched book on the social, economic, policy and environmental factors that are contributing to an obesity epidemic, still at the end, in the very last sentence, assigns 50 per cent of the responsibility to the individual, and individuals as a collective receive the other half of the blame. I see the assignment of responsibility to the individual as one of the major barriers to the resolution and correction of rising obesity levels. If responsibility is placed onto individuals and society as a collective of individuals, where is the onus and responsibility of institutions and governments to make progressive changes to rectify the social environment?

If continuing responsibility is placed on the individual the correct supportive structures must be provided by the government in order to ensure that people are able to make healthy choices. The growing policy concern over healthy eating has not reflected the specific issues facing low-income groups. Food poverty can be defined as 'the inability to access a nutritionally adequate diet and the related impacts on health, culture and social participation' (Friel and Conlon 2004: 1). This results in the precipitation of a power struggle within nutritional and health education in disadvantaged groups due to their lack of choice in eating habits as a consequence of food poverty.

This lack of choice and the inability to make what are framed as 'rational', 'correct' and 'healthy' decisions within health education can lead to attempts at expressions of power by the 'disadvantaged' subject through the rejection of knowledge produced by nutritional science. The education and dissemination of useful nutritional information is often hindered through the play of power associated with these types of knowledge. This knowledge is not being acknowledged, accepted and utilised because its authoritative position is an opportunity to reject a larger discourse of the subservient knowledge and experience of the disadvantaged subject. 'Current nutritional promotion is also criticised because it "lacks a social perspective and compassion" (Crotty 1995: 1) in that it fails to take into consideration the everyday realities of life which are believed to inform food decision making for most people.' (Coveney 1998: 463). The individualising nature of the rhetoric acts to problematise subjects' view points as we have seen in the cases of both Brenda and Johanna as they struggle to claim power through their nutritional knowledge. The experts stance on nutrition narrows the field to that of right and wrong and does not make allowances for the wealth and diversity of nutritional principles and definitions of health that are present in everyday discourse.

For nutritional education to be effective it must incorporate the discourses of the subjects on which it acts upon. While socially disadvantaged groups display awareness of what constitutes healthy eating, deciding on what to eat is a combination of factors and influences including personal and family preferences, attitudes, nutritional knowledge, social norms, access to shops and financial constraints. Without the realization of these factors and constraints and the acknowledgment of the subjects' experience, the education cannot succeed in its objective to introduce healthy eating behaviours. Even with the involvement of the subjects, changes need to be made to the physical environment that they reside in. It is not enough to educate people as to what is the 'healthy' choice. Additional steps must be made to create an environment that enables people to choose, one that gives them a choice.

References

Bartley, B. 1999a. 'Clondalkin APC 1998 Baseline Data Report: A Summary Interpretation'. Produced for the Clondalkin Area Partnership by the Land Use and Transportation Unit and Department of Geography, NUI Maynooth.

───── 1999b. 'Spatial Planning and Poverty in North Clondalkin', in M. Hennessey, D.G. Pringle and J. Walsh (eds), *Poor People, Poor Places: The Geography of Poverty and Deprivation in Ireland*. Dublin: Oak Tree Press, pp. 225–63.

Bourdieu, P. 'The Forms of Capital: English version published 1986', in J.G. Richardson (ed.), *Handbook for Theory and Research for the Sociology of Education*. New York: Greenwood Press, pp. 241–58.

───── 1977. *Outline of a Theory of Practice*. Cambridge: Cambridge University Press.

Coveney, J. 1998. 'The Government and Ethics of Health Promotion: The Importance of Michel Foucault', *Health Education Research* 13(3): 459–68.

Critser, G. 2003. *Fat Land: How Americans became the Fattest People in the World*. Boston, MA: Houghton Mifflin Company.

Davis-Floyd, R.E. and C.F. Sargent. 1997. *Childbirth and Authoritative Knowledge – Cross-Cultural Perspectives*. Berkeley, Los Angeles CA/London: University of California Press.

Department of Health and Children. 2000. *National Health Promotion Strategy 2000–2005*. Dublin: Department of Health and Children.

───── 2005. *Obesity: The Policy Challenges. The Report by the National Taskforce on Obesity*. Dublin: Department of Health and Children.

Department of Health. 1995. *A Health Promotion Strategy: Making the Healthier Choice the Easier Choice*. Dublin: Department of Health.

Foucault, M. 1990. *The History of Sexuality, Volume 1: An Introduction*. London: Penguin.

French, C. 2007. 'Britain May Take Custody of Obese Boy', The Associated Press; retrieved 26 February 2007 from: http://www.examiner.com/a586725~Britain_May_Take_Custody_of_Obese_Boy.html?cid=rss-World.

Friel, S. and C. Conlon. 2004a. 'What is the Extent of Food Poverty in Ireland?', *European Journal of Public Health* 13(4): 133.

───── and C. Conlon. 2004b. *Food Poverty and Policy*. Dublin: Combat Poverty Agency.

————, O. Walsh and D. McCarthy. 2004. 'The Financial Cost of Health: Eating in the Republic of Ireland', *Combat Poverty Agency Research Working Papers Series* at: http://www.combatpoverty.ie/publications/workingpapers/2004- 01_WP_TheFinancialCostOfHealthyEatingInIreland.pdf

Gastaldo, D. 1997. 'Is Health Education Good for You? Re-thinking Health Education through the Concept of Bio-power', in A. Petersen and R. Bunton (eds), *Foucault, Health and Medicine*. New York/London: Routledge, pp. 113–33.

Giddens, A. 1991. *Modernity and Self-Identity: Self and Society in the Late Modern Age*. Stanford, CA: Stanford University Press.

Petersen, A. 1997. 'Risk, Governance and the New Public Health', in A. Petersen and R. Bunton (eds), *Foucault, Health and Medicine*. New York/London: Routledge, pp. 189–206.

Saris, A.J., et al. 2002. 'Culture and the State: Institutionalizing "The Underclass" in the New Ireland', *City*, 6(2): 173–91.

Chapter 9

THE MAHARAJA MAC:

CHANGING DIETARY PATTERNS IN INDIA

Devi Sridhar

Introduction

Obesity rates have increased dramatically worldwide. In fact, it can be argued that rates are increasing most dramatically in developing countries from a very low baseline (Nishida and Mucavele 2005; de Onis 2005). The term developing country, taken from the World Bank, refers to low- and middle-income economies: technically those with a Gross National Product (GNP) per capita of less than $3,465. Developing countries have long been associated with chronic under-nutrition and infectious disease. However, in many developing countries there has been a concurrent increase in both obesity and under-nutrition often referred to as the 'dual burden' of malnutrition (Gillespie and Haddad 2003; Guntupalli 2005).

Economic development and other factors have led to changing patterns of living that can be viewed as part of the nutrition transition (Popkin 2004). Nutrition transition is generally defined as the shift away from a diet high in fibre and common carbohydrates towards more energy-dense diets that are high in sugars, refined foods, and saturated animal fats as well as a move towards a more sedentary lifestyle (Shetty 2002).

This chapter describes the process of this transition in India. It argues that there is a growing aspiration for, and move towards, consumption of Western food products in India. This can be viewed partially as a result of clever marketing practices in an environment of changing Indian sociocultural values and sustained economic liberalisation. India is an important country to examine since it is in the midst of an epidemiological shift manifested by the dual burden of both chronic and infectious disease. However, unlike other developing countries such as Mexico, Brazil and Chile, India has not yet exhibited an increase in the prevalence of overweight and obesity among the urban poor. Furthermore, while these other countries shown an inverted relationship between social class and obesity, this pattern is not yet observed in India. However economic development and other factors have led to changing patterns of living, which can be viewed as part of India's nutrition transition.

   This chapter is organized into four parts, starting with a brief discussion on economic liberalisation and urbanisation, turning to branding and then to the marketing practices of transnational companies. The last section is a case study of McDonald's in India which is used to illustrate the arguments. This chapter relies upon unpublished literature in India, fieldwork in the slums of New Delhi in September 2005 consisting of interviews with women, as well as secondary sources.

Globalisation and economic liberalisation

During a trip in 2005 to New Delhi, I picked up a copy of *Frontline* magazine. The cover article described the destitution, and severe under-nutrition, of the tribal communities in Maharashtra. That same day I attended a presentation on cholesterol at the India International Centre. The presentation was by an official from the World Health Organization (WHO) and was sponsored by the Almond Association in honour of World Heart Day. As the anecdote above shows, those living in India are concerned with the dual burden of under- and over-nutrition. A 2006 WHO study of employed Indian adults shows that the prevalence of individuals who are overweight in urban areas is high. It was estimated that 30.9 per cent of men and women were overweight (Reddy 2006). Guntupalli (this volume) notes that the NFHS-2 (National Family Health Survey (second survey conducted in 1998 and 1999)) data indicate that about 25 per cent of urban women are either overweight or obese. India already has the largest number of individuals suffering from diabetes, a co-morbidity of obesity, 19.3 million in

1995 projected to increase to 57.2 million in 2025. Even in poor areas, this is a problem. A 2001 study in the urban slums in Delhi found high rates of diabetes among dwellers (Misra et al. 2001).

Along with the rising rates of overweight and obesity, there exist huge rates of under-nutrition. India alone has 204 million individuals who are facing starvation on a daily basis. Guntupalli (this volume) notes that more than one-third of women aged fifteen to forty-nine have a BMI (Body Mass Index) less than 18.5 kg/m² indicating under-nutrition. Under-nutrition and poverty are concentrated among the historically marginalised groups such as scheduled castes and tribes, the elderly, women and individuals with disabilities. Under-nutrition is a manifestation of several underlying factors such as infection from an unsafe water source or unhygienic environment, lack of food security, intrahousehold gender inequality, and poor caring practices. As of 2004, an estimated 43.7 per cent of South Asian children were classified as stunted, stunted referring to being more than two standard deviations under the height-for-age National Centre for Health Statistics (NCHS) reference (UNDP 2003).

In contrast, diet-related diseases such as obesity, diabetes, cardiovascular disease and stroke have been traditionally viewed as diseases of the upper classes, simply confined to elite groups in India. High levels of obesity are attributed to lifestyle such as an unhealthy diet (for example; ghee, cream, dairy, oil) and inactivity, as well as a genetic predisposition: sometimes referred to as the thrifty genotype. Despite the high prevalence of overweight in the upper class, in the general population, as of 2000, only 2.1 per cent of children were classified as overweight. Thus over-nutrition is not yet a public health concern comparable to under-nutrition in terms of scale.

However, this is changing. In the past fifteen years – especially since India's economic liberalisation in the early 1990s – there have been rapid changes in diet and lifestyle among almost all sectors of the population (Reddy et al. 2005). Globalisation and its consequences in the economic, regional, social and cultural spheres can be viewed as responsible.

Economically, globalisation has unfolded in India driven by the new economic policies of liberalisation initiated in the early 1990s. These policies include reducing tariffs and duties on imported goods, opening up state controlled industries to the private sector and transforming the banking and investment sector. In terms of economic development, between 1991 and 2004, India's GDP has grown on average 6.2 per cent per year making it a fast growing economy it is now estimated that 300 million individuals can be viewed as middle-class (Table 9.1). Both in terms of demand (higher income on the part of individuals) and supply (greater access to market), the opening of India's economy has resulted in changes in dietary choice. The process of liberalisation is reflected in the public acceptance of India's transition to an open free-market economy indicated by new patterns of commodity consumption (Fernandes 2000).

In addition, globalisation has stimulated internal migration and urbanisation. It can be argued that urban localities are where dietary patterns are changing the fastest. Urbanisation has been noted as an important factor to explain the rising number of individuals who are overweight or obese (Shetty 2002), and it is argued that an increase in urbanisation leads to changes in diet regarding fat content of food, animal products, sugar and polished grains along with the adoption of a more sedentary lifestyle, compared with rural areas (Popkin 2004). In India, the urban population is growing faster than the rural population. Between 1991 to 2001, the population grew by 18 per cent in rural areas and 31 per cent in urban regions (Figure 9.1). The proportion of people in urban residence – presently around 30 per cent – is expected to rise to about 43 per cent in 2021 (Shetty 2002). Part of the explanation for the move towards urban living is the greater dependence on cash income for food and non-food purchases (Popkin 2004).

Economic transition and the move towards urban settings are changing the patterns of living in ways that increase the behavioural and biological risk factor levels for obesity in poor communities. Transnational companies have taken advantage of the new environment to launch targeted marketing campaigns aimed at branding.

Table 9.1: *Distribution of households by income (urban)*

Annual Real Income (rupees)	Income Class	1992–93 (%)	1995–96 (%)	1998–99 (%)
<35,000	Lower	38.4	27.9	19.0
35,001–70,000	Lower middle	33.0	34.9	33.8
70,001–105,000	Middle	16.1	20.3	22.6
105,001–140,000	Upper middle	7.6	9.6	12.2
>140,000	High	4.9	7.3	12.5

Source: Kim (forthcoming from NCAER 2003).

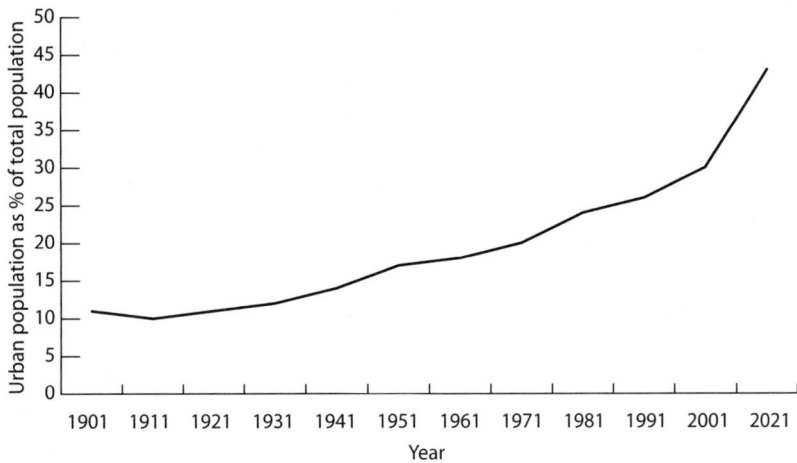

Figure 9.1: *Urbanisation in India, 1901–2021*
Source: Census 1991; Reddy et al. 1995.

Branding and the mass media

It has been shown that the Indian diet is shifting away from traditional foods (for example rice, ragi, wheat, lentils, vegetables) towards high-energy density, high fat and low fibre diets (e.g., animal products). If taken in sufficient quantity and diversity, the traditional diet is extremely healthy, especially when fish or chicken is included. For example, pulses and dried fish are comparatively cheap sources of protein.

At the national level this shift has been documented by quantitative data showing large increases in intake of egg, dairy and milk products along with edible oil. Food balance data from the Food and Agricultural Organization of the United Nations (FAO) show that per capita daily supply of animal products has increased from 7.0g in 1965 to 12.9g in 1999, thus contributing almost twice the energy content. The data also show a shift away from socially inferior cereal crops such as ragi, towards superior ones such as rice and wheat. In addition, the consumption of eggs and dairy products, which are important sources of saturated fat, has increased. This national trend can be viewed as representing a marked shift in the structure of the diet. India has shifted its energy distribution away from the diet dominated by complex carbohydrates to one with increased fats, added sugars and protein. The changing structure of the diet has both positive and negative health effects. It is positive because dietary diversity is increasing which reduces the occurrence of under-nutrition and micronutrient deficiencies. However the increased consumption of edible oil, added sugar, and animal products is linked with rapid increases in consumption of saturated fats and may displace essential nutrients, leading to increased positive imbalance and obesity.

When exploring the issue of diet, I asked women in my study area about their usual diet. One woman's response was telling. She said, 'All we have is rice, chappati, dahl, we cook with mustard oil, we're not rich. We have vegetables: yellow gourd, spinach, potato, tomato, pumpkin, but we cannot afford a healthy diet.' When I pushed as to what she viewed was healthy, she said, 'we cannot afford sweets, ghee, milk, Pepsi', and added, 'I worry about my children's health, about them not getting enough energy.' It might seem surprising that she was aware of Pepsi, but there was fairly widespread knowledge of drinks such as Pepsi, Fanta, Limca, Thumbs Up and companies such as McDonald's and Pizza Hut. Several other women repeated their desire for these goods as well. The simple explanation is that poor women want to have the products that the wealthy have. This is not a new observation. Rather, the interesting phenomenon was the widespread knowledge of different brands of Western foods and products and the belief that Western foods and products were healthier and better.

To illustrate how rapidly the branding has occurred, it is worth noting that Coca-Cola opened in India in 1993 and McDonald's in 1996. In thirteen or so years, these companies have reached segments of the population whom the government has been trying to reach in a concerted public health campaign for over fifty years with messages such as boil your water, have fewer children, use iodised salt and have children immunised.

Unlike the middle classes of India who are able to consume these Western products and eat outside the home, the slum community does not yet have the disposable income. Rather the aspiration for certain types of food exists. Among women who are illiterate, there is knowledge of Western brands and an association with these brands. As an employee of a marketing company said, 'We are dream merchants basically … we are writing to a [people] that cannot easily afford these things but who'd like to' (Fernandes 2000). The resulting question is why do women want to consume these foods? How has the association between Western products and health been created?

In Western countries, it has been argued that the consumption of processed and fast foods is a direct result of their affordability and palatability. The political-economic explanation for the occurrence of obesity in high-income countries such as the U.K, is that poor people are forced to eat energy-dense processed foods due to monetary, temporal or structural constraints. In addition, individuals are limited in their physical activity because they rely on a car or public transport to move around and have few outdoor, open spaces. Especially in poorer communities, parks, playgrounds and other outdoor areas are often not safe so individuals do not have opportunities for physical activity.

This is not the case in the poorer communities in Delhi. For example, during fieldwork, although women said that they would like to eat sweets and more energy-dense products such as dairy or meats, they did not have access to these due to cost constraints: energy-dense products are very expensive. Traditional diet items such as green leafy vegetables, spinach, lentils, and rice were not considered healthy foods or nourishing goods. The first explanation to be explored is that there exists the perception that those who are wealthy, those appearing on television, those who live in high-rises in the city, are not consuming the traditional diet, thus the traditional diet diminishes in status. Second, there is the general pull of Westernisation, especially in a city as international as Delhi. It was surprising that many women, despite not having a formal education, hold strong opinions about the world. For example, one of them asked me where I was from, when I replied England, she said, 'Oh yes, England is where they do not take baths and wear skimpy clothes.'

She had learned about England from television, a ubiquitous presence in urban settings. The influence of televison is pervasive. Almost every household I visited owned one or had access to one. Soap operas and advertisements are watched daily by families, especially children. It has been shown that many individuals cannot distinguish between programming and television. In fact, televison has been a negative factor in primary education efforts. One teacher who works in a state school attended by many Dalit (low-caste) children complained that the unregulated televison in the household was causing many of the children not to complete assignments and to do poorly in school.

Televison also contributes to reduced activity levels. In this particular field site, over 90 per cent of women do not work outside the home. Thus, their days are occupied by housework, caring for children, visiting neighbours, cooking and watching televison. This is viewed as a healthy lifestyle, perhaps since it is a wealthy lifestyle. When I asked one Muslim woman why she does not work and why she is not educating her teenage daughter, she said, 'When I ask my husband to work he says "Do I not earn enough?" ' Thus, not having to do manual labour or physically exert oneself is viewed as a sign of being better-off. While I was in the slums, I entered a tiny home. There was a bed, a little stove on the side of the house and a medium-sized televison. The woman of the household said that they had saved for many months to buy it. Physically and symbolically it was in the middle of the house and viewed as their prized possession. Thus, televison plays a central role in even poor communities.

As much anthropological literature has discussed, food is a medium for the expression of relationships, intimacy, distance, cooperation or conflict (Harris and Hoffenberg 1994). For example, 'rich' foods, rich both nutritionally and financially, have come to be a sign of social status and love (see Nandy 2004 for discussion of food as a social ritual). When hosting a party, it is necessary to provide rich foods because they are expensive. A meal not rich enough is a sign of either disregard for the guest or financial constraints on the part of the host. When attending a party, social rules dictate that a guest eats heartily, although there are accepted explanations for not eating such as having a 'sugar problem' (i.e., diabetes) or an upset stomach. Among the upper class, rich food is often viewed as healthy food. For example, one elderly woman remarked to me that *jangri*, an Indian sweet, is good for my health. *Jangri* consists of lentil and sugar balls deep fried in oil. Nutritionally, it is not healthy yet she classified it as a healthy food. Thus what is healthy is often associated with what is rich. An interesting area that still needs to be explored in India is the individual construction of healthy and unhealthy foods.

However, I should note that this is changing among the upper classes who are facing a diabetes epidemic. As a doctor in India remarked, 'I say you haven't made it in society until you get a touch of diabetes' (Kleinfield 2006). As the media attention to chronic diet-related disease increases, there is a general awareness of lowering intake of cholesterol, saturated fats, sugar and sodium. But because of the social aspects of food, there remains a tension between the public health messages to slim down and eat healthy with the social context of needing to provide and consume rich products. There is a conflicting statement: eat richer but weigh less. Adding to the confusion, there is a lack of clarity over what is in a healthy diet. Notions of what foods are healthy are often individually constructed and not based on scientific recommendations. There is also the general acceptance that obesity and chronic disease are inevitable, just another part of life. One gets old, and one gets sick. A friend who is twenty-nine and works as a reporter in Delhi recently emailed, 'I think we youngsters should also plan a Himalayan expedition (adventure cum spiritual) before we get really old and all the ailments strike us: BP [blood pressure], diabetes, hypertension to name a few.'

Marketing and consumerism

Overweight and obesity among the upper classes are not the issues this paper is addressing. Rather, it should be stressed that the importance is not necessarily what the wealthy are eating but instead the projection and subsequent perception of what the wealthy eat by the media to the middle and lower classes. Marketing campaigns of transnational companies focus on projecting an image to the middle and lower classes regarding a wealthier individual or household's consumption. The media projects an idea that what the wealthy do is Western. If a product is labelled in English or associated with being American, if it is expensive, consumers are satisfied to pay more regardless, because there is a belief that due to its cost, it must be a good product, and healthy. Having a label in English gives the product legitimacy. The perception exists that wealthy individuals eat outside the house. This creates the desire to eat non-traditional foods, and can be viewed as part of the general aspiration for a better life. As a reporter remarked, 'Make good money and get cars, get houses, get servants, get meals, get diabetes'(Kleinfield 2006). Last year, the MW fast-food and ice-cream restaurant in Chennai had a special offer, 'Overweight? Congratulations' (ibid). The promotion offered consumers discounts equivalent to 50 per cent of their weight in kilograms. The heaviest diner was 135 kg and ate for 67.5 per cent off.

In India, a changing sociocultural value system is indicated by the rising emphasis placed on materialism and consumerism as symbols of success (Panini 1996, cited in Kim forthcoming). Kim discusses how the new materialism encourages the attainment and display of consumer goods, which are valued 'not only because they assist individuals in dealing with the demands of daily life, but also because of the prestige derived from them' (Kim forthcoming). Higher classes determine social attitudes towards what is valued for consumption. As an editor of a print magazine said:

In the 1960s and 1970s this whole bit of accumulation of wealth was still suffering from a Gandhian hangover. Even though there were a whole lot of families who were wealthy all over India in the North and South if you noticed their lifestyles were very low key. There were not exhibitionist or they were not into the whole consumer culture. Now I see that changed completely ... You want to spend on your lifestyle. You want your cellphone. You want your second holiday home which earlier I said people would feel that sense of guilt- that in a nation like this a kind of vulgar exhibition of wealth is contradictory to Indian values. I think now consumerism has become an Indian value. (Fernandes 2000)

The marketing practices of transnational corporations such as McDonald's, Pizza Hut, Nic-Nac Fast Food and Coca-Cola have taken advantage of this association. By promoting the image of wealthy individuals consuming certain products, they are able to project an idea of what foods are healthy and what consumption is desirable. These companies have taken a major hit in the U.S.A. and U.K. for being unhealthy and uncool, and thus have reinvented themselves in India.

Television is arguably the most dominant gateway of globalisation affecting India. Advertisements using major Bollywood stars have been particularly effective at branding. Celebrities have continued to be associated with Western consumption. A Pepsi in hand has become a status symbol for the upwardly mobile. It can be argued that the media plays a major role in shaping dietary aspirations. The media is the dominant source of definitions of social reality for both individuals and groups and influences individual's consumption behaviour. To quote Fernandes, 'Idealized images of the urban middle class in the print media and television contribute to the production of images of an affluent customer, who has finally achieved the ability to exercise choice through consumption' (Fernandes 2000 in Kim forthcoming). These images influence the consumption aspirations of middle class workers, who seek to attain the lifestyle of the 'new rich', for example, information technology workers. The images are so powerful that even when certain individuals are unable to afford the high standard of living, they still perceive the consumption of the higher classes as beneficial. The media influences dietary aspirations through the images it offers of an ideal urban upper-middle class lifestyle (Liechty 2002). These are internalized by media consumers and used by them to construct the meaning of wealthy and poor lifestyles. A good example is that of

McDonald's, the fast-food restaurant chain. The following section relies on primary literature and Dash (2005).

Case study: McDonald's

McDonald's has effectively penetrated the upper and middle classes of India evidenced through its increasing sales revenue. McDonald's was one of the first foreign chains to move to India; its first store was opened in 1996 in Bombay. Onlookers noted that the store had mile-long lines and hundreds of people fighting for burgers and fries. India now has ninety-two McDonald's with eighteen based in New Delhi: which is not so many compared to its global reach. However, a McDonald's press release noted that India is viewed as the company's most promising new market. Because of the sensitive market, McDonald's has tailored its menu to India. There is no beef on the menu. There are no eggs in the mayonnaise. There are two separate burger cooking lines, one for vegetarians and one for meat. The workers in the vegetarian line wear green aprons and the workers in the non-vegetarian section are forbidden to cross over without showering first. In addition the menu has been tailored for India. For example, the Big Mac can be seen as McDonald's core product. In India, this has been replaced by the Maharaja Mac, which is a chicken burger. The rationale between the new burger is that as Hindus avoid beef and practising Muslims do not eat pork, only mutton and chicken satisfy the religious restrictions. Given that an overwhelming percentage of Indians (about 83 per cent) do not eat beef or pork, the introduction of the Maharaja Mac is the appropriate cultural fit. These hybrid products represent the ability of a multinational company to combine the national and the global within a singular narrative of hybrid commodity. The hybrid product can be viewed as the interlacing of the modern and the traditional. These products suggest that eating at McDonald's will not disrupt the stability of traditional Indian diet. The core of Indian tradition can be retained as the material context of the tradition is modernised and improved. Transnational companies are making explicit linkages between the traditional value of their products, the middle classes, who signify the wealth of the Indian nation, and the global.

In terms of pricing, McDonald's has employed a ladder strategy with prices starting at twenty-nine rupees, going up to eighty-nine rupees. This broad range of prices for value meals is to ensure that most sections of customers are catered for in terms of affordability. As a McDonald's Managing Director in India said, 'Our clear strategy is to bring the customers in initially and to provide a range of entry-level products so that they can try new items and graduate to the higher range'. While these prices might seem low compared to international standards, their price relative to local foods is expensive. McDonald's has created itself as a luxury good and targets a different demographic population in India, the upper and upper middle-class, compared to its image in North America and Western Europe. In India, McDonald's delivers through its service McDelivery. McDonald's also caters for parties. The provision of services aimed at the upper classes can be viewed as brand creation. While McDonald's is viewed as a fast food in the U.K., meaning it is cheap, affordable and unhealthy, in India it has come to symbolize wealth, excess, and being part of an elite. In addition as noted earlier with reference to labelling goods, because it is both expensive and American, it is perceived as healthy and good for children.

The emphasis on children results in another issue: McDonald's is tailored towards children in India. Outlets are called 'McDonald's Family Restaurants', unlike McDonald's in other parts of the world. McDonald's restaurants provide a clean, comfortable and stress-free environment. Especially important in India's hot climate is that most outlets are air-conditioned. Kishore Dash has called this the 'McDonald's experience', described as eating in a clean, friendly and fun-filled environment with quick and accurate services (2005). In addition, as one woman noted, 'What attracts me to McDonald's? The ability to give my children a slice of America.' Thus McDonald's also has come to be associated with certain American traits such as success, productivity and a good life. This growing acceptance corresponds to the big impact of the American influence on Indian entertainment, business, and diet.

McDonald's branding strategy in India is different to its brand creation in the U.S.A. The company is using a different marketing strategy tailored to the new ideas of the Indian Dream, targeting the aspirations of the lower and lower middle-class. McDonald's is entering the Indian market with a long-term branding strategy. Since 1997, McDonald's has had an annual revenue growth of 50 per cent. With India's population already over 1.1 billion, the potential customer base for McDonald's is larger than the size of entire developed countries. Thus the marketing and branding seems also to be targeted at the vast majority who are still too poor to consume these goods but yet have access to media and aspire to have these goods. As the owner of a Subway in New Delhi said, 'With the number of working households growing and the income levels increasing, a fast food revolution is around the corner.'

Conclusion

This paper has argued that there is a growing aspiration for and move towards consumption of Western food products in India. This can be viewed partially as a result of clever marketing practices in an environment of changing Indian sociocultural values and sustained economic liberalisation. As India's economy grows and incomes increase, there is a greater likelihood that animal products and 'rich foods' will be consumed. In development, the effects of increased income have generally been viewed as beneficial, since higher income is associated with 'better quality diets, better health care, better child growth, and lower morbidity and mortality from infectious disease' (Du et al. 2004). Concurrently, income increases might induce negative dietary changes such as increased consumption of animal and processed foods. McMichael (2004) notes:

> Diets have a political history framed by class, cultural and imperial relations. Animal protein consumption signals rising affluence and emulation of Western diets. Movement up the food chain hierarchy, from starch, to grain, to animal protein and vegetables, is identified with modernity.

Increasing rates of obesity in India will result in dramatic increases in chronic disease such as diabetes, hypertension, and cardiovascular disease as well as reproductive conditions. Obese women have a higher risk of complications during pregnancy such as hypertensive diagnosis and gestational diabetes and also have delivery complications such as high rates of caesarean and prolonged delivery. Obese women also have a higher risk of miscarriage.

The coexistence of under-nutrition and overweight in the same country, region and even household, particularly in urban areas, is an important issue and will become more apparent as an increasing number of lower class households shift towards a more energy-dense and low fibre diet, and lower activity levels. The speed of the nutrition transition in India is increasing the likelihood that both problems, under-nutrition and obesity, will coexist. Thus, under-nutrition and obesity are not opposing health concerns but rather a complex phenomenon that must be addressed using integrated strategies. These strategies must take into account how dietary patterns are changing in India and are likely to change, the impact these will have on public health and integrate these concerns into public health interventions.

References

Dash, K. 2005. 'McDonald's India', Thunderbird Case Study A07-05-0015, accessed 15 September 2006 at: http://www.thunderbird.edu/about_thunderbird/case_series/2005/_05-0015.htm.

De Onis, M. 2005. 'The Use of Anthropometry in the Prevention of Childhood Overweight and Obesity', *United Nations Standing Committee on Nutrition* 29: 5–12.

Du, S., et al. 2004. 'Rapid Income Growth Adversely Affects Diet Quality in China – Particularly for the Poor!', *Social Science and Medicine* 59: 1505–15.

Fernandes, L. 2000. 'Nationalizing "the Global": Media Images, Cultural Politics and the Middle Class in India', *Media, Culture and Society* 22: 611–28.

Gillespie, S., and L. Haddad. 2003. *The Double Burden of Malnutrition in Asia: Causes, Consequences and Solutions.* New Delhi: Sage.

Guntupalli, A. 2003. 'Inquiry into the Simultaneous Existence of Malnutrition and Obesity in India', Seminar at the University of Bristol, 13 April.

Harris-White, B. and R. Hoffenberg (eds). 1994. *Food: Multi-Disciplinary Perspectives.* Oxford: Blackwells.

Kim, Y. Forthcoming. 'The Inner and Outer Meanings of Alcohol in India'. D.Phil. dissertation. University of Oxford: Queen Elizabeth House.

Kleinfield, N.R. 2006. 'Modern Ways Open India's Doors to Diabetes', *New York Times*, 13 September.

Liechty, M. 2002. *Suitably Modern.* Princeton: Princeton University Press.

McMichael, P. 2004. *Development and Social Change: A Global Perspective.* London: Sage.

Misra, A., et al. 2001. 'High Prevalence of Diabetes, Obesity and Dyslipidaemia in Urban Slum Population in Northern India', *Nature* 25: 1722–9.

Nandy, A. 2004. 'The Changing Popular Culture of Indian Food: Preliminary Notes', *South Asia Research* 241: 9–19.

Nishida, C., and P. Mucavele. 2005. 'Monitoring the Rapidly Emerging Public Health Problem of Overweight and Obesity: The WHO Global Database on Body Mass Index', *United Nations System Standing Committee on Nutrition* 29: 5–12.

Popkin, B.M. 1998. 'The Nutrition Transition and its Health Implications in Lower-income Countries', *Public Health Nutrition* 1: 5–21.

———— 2004. 'The Nutrition Transition: An Overview of World Patterns of Change', *Nutrition Reviews* 62: S140–43.

Reddy, K.S., et al. 2005. 'Responding to the Threat of Chronic Diseases in India', *The Lancet* 366(9498): 1744–49.

————, et al. 2006. 'Methods for Establishing a Surveillance System for Cardiovascular Diseases in Indian Industrial Populations', *Bulletin of the World Health Organization* 84(6): 461–49.

Shetty, P. 2002. 'Nutrition transition in India', *Public Health Nutrition* 5(1A): 175–82.

United Nations Development Programme (UNDP). 2003. *Tamil Nadu: Human Development Report,* accessed 15 September 2006 at: http://hdr.undp.org/en/reports/nationalreports/asiathepacific/India/name,3254,en.html

Chapter 10

IS THERE A RELATION BETWEEN FATNESS AND REPRODUCTIVE HEALTH?

A STUDY OF BODY MASS INDEX AND THE REPRODUCTIVE HEALTH OF INDIAN WOMEN

Aravinda Meera Guntupalli

Introduction

Overweight and obesity are prevalent, not only in developed countries but also in developing countries, like India, which are experiencing nutritional transition. Diet-related diseases such as obesity, diabetes mellitus, cardiovascular disease, hypertension, and stroke are increasing in India owing to changes in dietary patterns and lifestyle (Anate et al. 1998). Rapid changes in dietary patterns and lifestyles resulting from industrialisation, urbanisation, economic development and globalisation have a significant impact on populations with nutritional transition, like that of India (Popkin 1993; Griffiths and Bentley 2001; Shetty 2002). However, the fact that obesity can lead to many reproductive health problems is less established. Recent clinical and biomedical research supports the relation between obesity of women and reproductive health problems like infertility, abortion and pregnancy complications (Jarow et al. 1993; Moran and Norman 2002). Interestingly, Hippocrates, a thousand years ago, documented the consequences of nutritional status and obesity (from Chadwick and Mann 1978). He recognised that 'sudden death is more common in those who are naturally fat than in the lean'. Moreover, he emphasised a consequence for reproduction: 'Fatness and Flabbiness are to blame. The womb is unable to receive semen and they [fat women] menstruate infrequently.'

However, until the International Conference on Population and Development (ICPD) in 1994, the reproductive health of women in general was not given much attention by demographers and anthropologists. Here, for the first time, the reproductive health of women was defined by the ICPD as a state of complete physical, mental and social well-being and not merely the absence of disease or infirmity, in all manners relating to the reproductive system and to its functions and processes (ICPD 1994).

This definition has undergone several modifications and the recent reproductive morbidity definition was further broadened due to its inclusion of a wide range of reproductive health problems. Currently, it encompasses obstetric morbidity, including conditions during pregnancy, delivery and the post-partum period, and gynaecological morbidity, including conditions of the reproductive tract not associated with a particular pregnancy, such as reproductive tract infections, cervical cell changes, prolapse and infertility. In addition, it also encompasses related morbidity such as urinary tract infections, anaemia, high blood pressure, obesity and syphilis, as systematic conditions (Zurayk et al. 1993). Finally, several aspects of reproductive health including infertility have started to receive attention from social scientists.

In India, until recent years, health planners focussed mainly on overpopulation. The main reproductive health issue was the reduction of the total fertility rate to control population growth and accordingly, most of the research was aimed only at fertility-related aspects. It is only in recent years that infertility has received more attention in India. According to NFHS-2, almost 4 per cent of Indian women aged forty to forty-four have not had any children, and 3.5 per cent of currently married women were defined as infertile (International Institute for Population Sciences (IIPS) and ORC Macro 2000). In general, sexually transmitted diseases (STD), maternal health factors, poor health, nutrition, lifestyle, availability and accessibility to reproductive health services were found to be correlated with infertility (Jejeebhoy 1998). The relation between STDs and infertility gained some attention due to recent HIV (Human Immunodeficiency Virus) and STD awareness programmes. However, the role of fatness in causing reproductive health problems is not established in India. In sum, infertility did not receive much attention and the role of fatness in deciding reproductive morbidity in India was further ignored. Hence, taking into consideration increasing obesity in India, the current study aims to look at the relation between obesity and reproductive health problems like infertility and abortion.

On one side, reproductive health problems seem to increase in India and on the other side India is undergoing rapid nutritional transition. We can clearly see the rapid transformation in dietary patterns using the calories consumption data over time. The amount of calories per person per day using Food and Agriculture Organization data, from 1967 to 2002, for India, shows an increase from 2,041 calories to 2,420 calories (FAO 2005). Although protein intake stagnated in this period, we see a rapid increase in the consumption of fat. In addition, urban residence seems to increase Body Mass Index (BMI), and my research shows that urban women have a higher probability of being overweight and obese compared to rural women in all standards of living categories (Guntupalli 2007a; Hussain 2007). Moreover, urban women with a low standard of living have higher odds of being overweight and obese compared with rural women with middle and low standards of living. This clearly indicates the fact that urban residence is a more important factor in deciding Body Mass Index than the income status.

In this study, the overall focus is on the relation between obesity and reproductive health problems in India using National Family Health Survey data. In the next section, I review the relation between obesity and reproductive health using the biological literature. Then I focus on data and methodological issues. After that, I discuss the prevalence of obesity in India and, finally, offer an analysis of the relation between reproductive health and obesity/overweight.

Relation between body fat and reproductive health

Many previous studies from social science disciplines considered obesity and infertility as two separate issues. Researchers working on infertility did not focus on obesity (Unisa 1999; Guntupalli and Chenchelgudem 2004) and researchers working on obesity did not take into consideration increases in reproductive health problems due to increasing body weight (Popkin 1993; Griffiths and Bentley 2001). Yet, recent biological research shows that these two issues are not exclusive (Norman and Clark 1998; Lake et al. 1997). Human reproduction is affected by both over- and under-nutrition. This intertwined relation of fatness and reproductive health needs attention not only from biologists but also from social scientists.

Biologists explored this relation in detail for more than a decade and the relation between obesity and infertility (or subfertility) has attracted attention among biologists since the last years of the twentieth century (Hamilton-Fairley et al. 1992; Crosignani et al. 1994; Norman and Clark 1998). Moreover, some recent epidemiological studies have shown that obese women have alterations in their menstrual cycle, chronic or intermittent anovulation, and excess androgens. For example, Yen (1999) showed that women with a BMI greater than 30 kg/m^2 have abnormalities in secretion of hypothalamic GnRH (Gonadotropin Releasing Hormone), pituitary LH (Luteinizing Hormone), and FSH (Follicle Stimulating Hormone) that result in anovulation. Moreover, a study by Lake et al. (1997) that used a large cohort of women concluded that menstrual disorders are frequent in girls that were obese during their adolescence compared to girls that were obese during their childhood.

If there is a relation between fatness and reproductive health, what are the mechanisms according to the recent biological work? Although the relation between body fat and reproductive health is clearly established in recent years, there is ambiguity about the mechanism through which body fat can influence fertility. Evidence supporting the interrelation between fertility and insulin has emerged recently (Hunter et al. 2004). However, the role of adipokines – including leptin, resistin and adiponectin – with obesity and insulin resistance is recently discussed by Kadowaki et al. (2003), and Greenfield and Campbell (2004).

Mitchell et al. (2005) have shown a clear link between obesity and infertility by discussing in detail the role of adipokines. They demonstrate that adipokines leptin, adiponectin and resistin produced by adipose tissue, are altered with obesity, which in turn influences not only energy homeostasis but also female fertility. Also, leptin influences the developing embryo, the functioning of the ovary and the endometrium, and interacts with the release and activity of gonadotrophins and the hormones that control their synthesis (Mitchell et al. 2005). In sum, obesity alters adipokines and this in turn results in reproductive morbidity.

Interestingly, like social scientists working on infertility, biological scientists focussed heavily on females while studying infertility and obesity. The relationship between obesity and infertility in males has not been given much attention in either Indian or international clinical and biomedical research. An exception to this is the recent study by Jarow et al. (1993), which focussed exclusively on male fertility and obesity. Jarow et al. studied the effect of obesity and fertility status on sex steroid levels in men, by comparing endocrinology of both fertile and infertile obese men, with that of both fertile and infertile non-obese men. Interestingly, Jarrow found that there was a significant sex steroid level difference in obese infertile compared to the remaining three groups. Serum Steroid Hormone Binding Globulin (SHBG) was significantly lower and this was found to be correlated with elevated bioavailability of both testosterone and estradiol in the obese infertile group. However, serum LH (Luteinizing Hormone) levels were not different, suggesting that free testosterone levels were unchanged. They concluded that reduction of serum SHBG, total testosterone, and testosterone/estradiol ratio appear to be a marker of infertility among obese men. Nevertheless, there is still very little research to support the mechanism in detail, or to understand what relation there may be between obesity and erectile dysfunction (Jarow 2006; Riedner et al. 2006).

As mentioned earlier, it is only recently that infertility has received more attention in India. Results of research on increasing infertility prevalence in India and its consequences pointed clearly that women are targeted as childbearers and are looked down upon if they are childless (Unisa 1999; Mulgaonkar 2001; Guntupalli and Chenchelgudem 2004). Moreover, most of these previous studies are based on small samples and specific states. My data reflect information based on the anthropometric and reproductive health data of nearly ninety thousand women, which can help us in understanding these complex relationships. The advantage of using large surveys is the representativeness of the sample by covering all the regions and socioeconomic groups. Moreover, the biological studies mentioned above are focussed on relatively small samples and by using a demographic survey it is possible to focus on biological and social aspects to establish relationships between body fatness and reproductive health.

Data and methodology

The data referred to are the NFHS-2 (International Institute for Population Sciences and ORC Macro 2000) that covered a representative sample from twenty-six states in India, of around 91,000 women aged fifteen to forty-nine, who had been married at least once. NFHS-2 was conducted in two phases: the first starting in

November 1998 and the second in March 1999. This survey provides state-level estimates of demographic and health parameters as well as data on various socioeconomic components. Using secondary-level data analysis, it is possible to estimate infertility indirectly. Furthermore, we can study the relation between the estimated infertility and the available BMI data. However, in this study, it is difficult to exactly locate the main reasons for infertility, unlike clinical studies.

To test the hypothesis that obesity is related to reproductive problems, I used information about current age, marital status, contraceptive usage, children born, pregnancy complications, duration of marriage, height, weight and other socioeconomic and demographic indicators of women. Although the survey was based on women aged fifteen to forty-nine, the data analysed in this study were limited to 'currently married' women aged between twenty and forty-four, to control for adolescent subfertility and menopause. The concept of 'adolescent sterility' is very important in this context, as it plays a defining role in women not becoming pregnant for a certain period of time and this period is extended with the increase in exposure to hunger and malnourishment (MacCormack 1982; Guntupalli and Chenchelgudem 2004). The NFHS-2 survey did not collect anthropometric data of women who were pregnant at the time of the survey, nor women who had given birth during the two months preceding the survey. This exclusion further reduced our observations. Secondary infertility was not taken into consideration as it is extremely complicated to capture it using indirect measurement.

Indirect estimation of infertility

Infertility was defined clinically and measured indirectly by several researchers. Some scholars have defined infertility as having no pregnancy after a year of unprotected intercourse (Thonneau and Spira 1990; Schmidt and Munster 1995) where others have defined it as no pregnancy after five years of marriage (Larsen 2003). However, the methodology of selecting infertile women in my study was rather strict compared to the previous studies due to various considerations as mentioned below.

(1) Women must experience a marital duration of five years.
(2) Women must be married at the time of the study. Ergo, no widowed or divorced women were included in the analysis.
(3) The definition of infertility excludes younger and older women to control for adolescent sterility and menopause. To be precise the definition was applied to women aged twenty to forty-four.

(4) Moreover, only non-contraceptive users were included in the sample.
(5) The relation between 'fatness' and infertility was studied by controlling for reproductive health problems of women, which allows measuring the relation between infertility and fatness precisely.

Larsen's (2003) infertility analysis, using secondary large-scale demographic health survey data, considered current users of contraception as fertile and this estimation has slightly lowered her estimates of infertility as some contraceptive users might not know their infertility status. On the contrary, it is not possible to consider the users of contraception as infertile women. Although my current selection would lead to a biased estimate of infertility, it does not alter our aim: to study the relationship between obesity and infertility. The major focus of our study is to find connections between obesity and reproductive health but not to construct the exact prevalence of infertility.

However, this indirect estimation method has both advantages and disadvantages. The major advantage is the large sample size, reliability and national coverage. Our restricted sample includes more than 26,000 women from all regions of India, so this sample is reliable for studying the impact of 'fatness' on infertility. Moreover, as the data are now readily available, indirect estimation of infertility using secondary data becomes affordable and quick. Also, using this data, it is possible to control for gynaecological morbidity based on the symptoms of the respondents, such as pain during urination, odour, pain while having intercourse or abdominal pain.

Regarding disadvantages, this study focusses exclusively on women and ignores men. By studying female infertility, this study neglects the obesity and reproductive health problems of the husbands of the respondents. Moreover, we have no information about the duration and prevalence of their reproductive health problems. It is not possible to assess the reasons of infertility using such secondary data. In addition, there might be underreporting of some reproductive health problems like abortion and infertility, due to the fear of being discriminated against. Infertility is not culturally accepted in a society like India. Only twenty-six thousand women were considered in this study due to the selection criteria although the original NFHS survey collected information on nearly ninety thousand individuals (IIPS and ORC Macro 2000). Hence, the final sample used in this study is heavily reduced by the selection of adult currently married non-contraceptive users that have recorded their Body Mass Index.

How does it differ compared to other studies? In this study, infertility was measured as 'absence of children' for 'currently married women' after 'five years of their marriage'. This study excluded the divorced and widowed women, women aged below twenty and women who used contraception during the survey. Moreover, for the first time women were tested for their childlessness by controlling their gynaecological morbidity. Using this data, I can assess non-clinically the prevalence of infertility among Indian women, as an indirect way of further studying its relationship with body fat.

Overweight and obesity in India: how serious is the problem?

Obesity and overweight prevalence in India is increasing rapidly and has been for the past decade. In this section, I will discuss the problem of overweight and obesity by analysing secondary data. The NFHS-2 published report shows that more than one-third of women aged fifteen to forty-nine have a BMI less than 18.5 kg/m2 and 25 per cent of urban women are either overweight or obese (IIPS and ORC Macro 2000). According to the World Health Organization (WHO) classification, nearly 11 per cent of Indian women belong to the overweight category. The percentage of women belonging to the overweight category is higher among women from the Sikh (26.2 per cent) and Jain (28.8 per cent) religions (IIPS and ORC Macro 2000).

How different is our sample compared to the total sample in terms of their Body Mass Index? It is very important to compare the original and the restricted samples to identify a potential bias in this study. The percentage of overweight and obese women in our sample is very low (Table 10.1) compared to the original data. This is mostly due to the exclusion of contraceptive users as it is likely that contraception is positively related to urban residence and a higher standard of living; hence body fatness.

What are the determinants of overweight and obesity? The prevalence of obesity and overweight is determined not only by the standard of living but also by the urban or rural place of residence. To unveil the complexity of this relation, a compound variable was constructed by taking into consideration standards of living and the respondents' place of residence.

Table 10.1: *Comparison of BMI of original sample and our study sample*

	BMI of total sample	BMI of our study sample
Malnourishment	32.94	35.00
Normal weight	48.91	57
Overweight	13.50	6.1
Obesity	4.65	1.5
Total	72,607	26,027

The total sample for anthropometric indicators is 77119, and from that too tall and too fat people were excluded.

Source: Computed from National family Health Survey–2.

Using the compound variable that has six categories, it is very clear (Table 10.2) that women with a high standard of living, especially from urban areas, have higher percentages of overweight and obesity. Urban women with a middle standard of living and rural women with a high standard of living experience similar overweight prevalence. Although standards of living play an important role in deciding fatness, we can conclude that that urban area of residence is the most important factor.

Table 10.2: *Percentage of women with overweight and obesity by standard of living (SLI) and place of residence (rural/urban)*

	Rural			Urban		
	Low SLI	Middle SLI	High SLI	Low SLI	Middle SLI	High SLI
Overweight	2.0	5.3	14.9	5.3	14.2	26.2
Obesity	0.3	0.9	3.6	1.1	3.5	9.8
Sample	21535	25375	7244	2530	9202	8655

SLI refers to Standard of living and it is a composite indicator based on household amenities in the household.

Source: Computed from National family Health Survey–2.

Apart from socioeconomic factors and area of residence, regional factors play an important role in the occurrence of obesity and overweight and the prevalence of overweight and obesity exhibits clear regional differences. Women from the north, west and the south have higher percentages of obesity and overweight compared to other, eastern, regions (Table 10.3).

Table 10.3: *Percentage of women with overweight and obesity by region*

	North	Central	East	North east	West	South
Overweight	12.3	6	5	4.6	10.2	12.2
Obesity	4.3	1.5	0.8	0.9	3.5	2.9

Regions are classified using National Family Health Survey geographical clustering.

Source: Computed from National family Health Survey–2.

In sum, both overweight and obesity have become substantial problems among several groups of women in India, particularly women living in urban areas, and women from households with a high standard of living. In the next section, I discuss increasing fatness in India and its relation to infertility and other reproductive health problems.

Reproductive health and fatness: can this relationship be confirmed for Indian women?

In the previous sections, biological literature about the relation between anthropometry and infertility was reviewed along with some analysis that reflected the situation of nutritional transition in India. Now, in this section we will move to the prevalence of infertility using the methodology that was discussed in the previous sections. Moreover, we will test the relation between BMI and reproductive morbidity to test various questions that were raised earlier.

Nearly 12.3 per cent of currently married women that did not use contraception during the survey had no children after at least five years of marriage. Multivariate analyses were conducted to determine the risk factors of primary infertility. Specifically, logistic regression was used to estimate the relative odds of being primarily infertile or childless after at least five years of marriage, controlling for the effects of obesity and gynaecological morbidity.

Before discussing the relationship between infertility, body fatness and reproductive health, it is worth studying the gynaecological morbidity in our study sample. NFHS-2 collected information from women on some common symptoms of gynaecological morbidity, like abnormal vaginal discharge, urinary tract infections in the three months preceding the survey, and intercourse-related pain.

Specifically, the prevalence of reproductive health problems among ever-married women is estimated from women's self-reported experiences with each of the following problems: irritation around the vaginal area, bad odour, severe lower abdominal pain, pain or burning while urinating, or painful intercourse. Women who experience one or more of these gynaecological health problems could either have or be at risk of getting a reproductive track infection or sexually transmitted diseases. However, since information on health problems is based on self-reported symptoms rather than clinical tests, the results should be interpreted with caution. Using self-reported symptom data we cannot precisely understand the duration of their morbidity. Moreover, we cannot reflect on the exact type of morbidity, severity of the problem and the significant impact of these symptoms on reproductive health problems like infertility. Nevertheless, despite the lack of precision, the self-reported gynaecological morbidity indicators can be used as a proxy due to unavailability of other indicators. Besides, the aim of this study is not to estimate the incidence of reproductive morbidity with precision.

Based on the study sample, more than 18 per cent had itching or irritation problems, almost 19 per cent had pain during urination and over 20 per cent had abdominal pain (Table 10.4). And notably, 40 per cent of the women seem to experience at least one of the symptoms. If sexually transmitted diseases or other reproductive morbidities are a major cause of infertility, it is very important to include at least an indirect indicator to control for its impact on infertility, to study the clear relation between obesity and reproductive health. The use of odds ratio from our analysis will show if the chances of being infertile increases by BMI: even after taking into consideration their gynaecological morbidity. If the odds ratio turns out to be more than one, it can be concluded that the chance of being infertile increases for other categories compared to the reference category. Moreover, if our BMI variable stays significant even after controlling for reproductive morbidity, we can conclude

Table 10.4: *Percentage of women reporting gynaecological morbidity symptoms during the three months preceding the survey*

Symptoms	Percentage
Itching or irritation	18.53
Bad odour	13.28
Abdominal pain	20.43
Pain or burning while urinating	18.85
Pain during intercourse	13.13
Any of the above	40.06

Source: Computed from National family Health Survey–2.

that fatness in women also plays an important role in deciding infertility along with reproductive morbidity.

Using logistic regression analysis (Table 10.5) it can be concluded that the probability of infertility increases for the obese and overweight women compared to normal and underweight women. Moreover, the odds of infertility increases for women with some reproductive morbidity symptoms compared to women that did not have any symptoms of reproductive morbidity during the survey.

Table 10.5: *Effect of BMI and reproductive morbidity on infertility*

Probability of occurrence of infertility	Odds ratio
Body Mass Index	
Overweight and Obesity	1.7*
Others (reference category)	
Reproductive morbidity	
Symptoms	1.35*
No symptoms (reference category)	

Only the variables with * have significant effect (p 0.05).

Source: Computed from National family Health Survey–2.

What about pregnancy complications, and abortion and the relation with BMI? The unrestricted data were tested for the odds of occurrence of reproductive health problems like abortion, and for caesarean surgery during delivery for overweight and obese women. It is very clear from this data that obese and overweight women have higher odds compared to malnourished women in having caesarean delivery. Concurrently, the odds of abortion have increased very much for obese and overweight women compared to the malnourished. Even after controlling for economic and regional variables, the relation between fatness and caesarean delivery stayed significant (Table 10.6). A similar argument can hold good for the relation between fatness and abortion (Table 10.6). We can conclude that fatness is a significant determinant of reproductive health problems: researchers working on the impact of obesity on diabetes and coronary heart diseases should also address reproductive health problems that arise with increasing BMI.

Table 10.6: *Effect of BMI socio/demographic factors and regions on reproductive health of women*

Region	Caesarean Delivery	Abortion
North (reference category)		
Middle	0.99	1.01
East	0.86	2.32*
North-east	1.71*	1.779*
West	1.01	1.468*
South	1.82*	2.614*
Body Mass Index		
Malnourished (reference category)		
Normal	0.98	1.28*
Overweight and Obesity	1.42*	2.01*
SLI (Standard of living) and Residence		
Rural low SLI (reference category)		
Rural middle SLI	0.75	2.14*
Rural high SLI	0.68	3.65*
Urban low SLI	0.74	1.42
Urban middle SLI	0.69	1.64*
Urban high SLI	0.59*	2.63*

Only the variables with * have significant effect (p 0.05).

Source: Computed from National family Health Survey–2.

Discussion

How can we use the findings in a broader way? It is very important to reflect on the current findings according to the current and predicted future trends of nutritional transition in India. India is simultaneously experiencing high levels of malnutrition and increasing levels of obesity (Shetty 2002; Guntupalli 2007b). Child stunting, especially in female children, is higher in India compared to many parts of the world (Guntupalli and Schwekendiek 2006). The high levels of childhood stunting will increase the obesity rates in the future, as it is argued that population with higher rates of stunting and low birth weight will have an increased risk of obesity related to chronic diseases in adulthood when exposed to nutritional transition (Byers and Marshall 1995; Scrimshaw 1995; Barker 1998; Darnton-Hill and Coyne 1998). However, a recent study by Guntupalli (2007a) using the 1998 NFHS-2 data showed that taller women were prone to obesity: but the relation might change in the later phases of nutritional transition. In later phases of nutritional transition, it is possible that the relation between childhood stunting and adult obesity becomes robust. If this happens, it will be a huge future challenge for Indian policymakers, for children that are stunted would be overweight in their adulthood. Consequently, their reproductive health would be negatively affected, too.

Conclusion

I conclude that there is a significant relation between obesity and reproductive health problems in India. Using national data and advanced methods, I could establish a clear relation between body weight and reproductive health problems like abortion, infertility and caesarean delivery even after controlling for geographic location and economic status. So what are the implications based on this study? India's nutritional transition is in the beginning phase, and in later stages, there will be an increase in the percentage of overweight and obese women. Hence, Indian women will not only face the problem of increasing body weight but also reproductive health problems as per my analysis. To avoid this cause, we need to consider two issues. First, there is a need for further research to establish the relation between fatness and reproductive health using information of couples and better reproductive health indicators to improve the quality further. Second, the results from such research should be disseminated to the policymakers, media and risk groups. As a matter of fact, increasing fatness will not only have biological consequences like higher infertility but also gender and social related problems – like discrimination against infertile women and families, impact on marriage and domestic violence – that need to be addressed.

References

Anate, M., A.W.O. Olatinwo and A.P. Omestina. 1998. 'Obesity – an Overview', *West African Journal of Medicine* 17(4): 248–54.

Azziz, R. 1989. 'Reproductive Endocrinologic Alterations in Female Asymptomatic Obesity', *Fertility and Sterility* 52: 703–25.

Barker, D.J. 1998. *Mothers, Babies and Health in Later Life*. Edinburgh: Churchill Livingstone.

Byers, T. and J.A. Marshall. 1995. 'The Emergence of Chronic Diseases in Developing Countries', *SCN News* 13: 14–19.

Chadwick, J. and W.N. Mann. 1978. *Hippocratic Writings* (trans.). Harmondsworth: Penguin Books.

Crosignani, P.G., et al. 1994. 'Anthropometric Indicators and Response to Gonadotropin for Ovulation Induction', *Human Reproduction* 9: 420–23.

Darnton-Hill, I. and E.T. Coyne. 1998. 'Feast and Famine: Socioeconomic Disparities in Global Nutrition and Health', *Public Health Nutrition* 1(1): 23–31.

Food and Agriculture Organization (FAO) Statistics Division. *Food Security Statistics*, accessed November 2005 at: http://www.fao.org/es/ess/index_en.asp.

Greenfield, J.R. and L.V. Campbell. 2004. 'Insulin Resistance and Obesity', *Clinical Dermatology* 22: 289–95.

Griffiths, P.L and M.E. Bentley. 2001. 'The Nutrition Transition is underway in India', *Journal of Nutrition* 131: 2692–700.

Guntupalli, A.M. 2007a. 'Nutritional Transition in India: A Special Focus on Severe Malnutrition and Increasing Obesity', in A. Hussain (ed.), *Managing Overweight and Obesity*. Punjagutta: ICFAI University Press.

───── 2007b. 'Anthropometric Evidence of Indian Welfare and Inequality in the Twentieth Century', Ph.D. dissertation. University of Tübingen.

───── and D. Schwekendiek. 2006. 'A Worldwide Study on the Gender Differences in the Welfare of Children', Working Paper, Economics and Human Biology Conference, 2006, University of Tübingen.

───── and P. Chenchelgudem. 2004. 'Perceptions, Causes and Consequences of Infertility among the Chenchu Tribe of India', *Journal of Reproductive and Infant Psychology* 22(4): 249–59.

Hamilton-Fairley, D., et al. 1992. 'Association of Moderate Obesity with a Poor Pregnancy Outcome in Women with Polycystic Ovary Syndrome Treated with Low Dose Gonadotrophin', *British Journal of Obstetrics and Gynaecology* 99: 128–31.

Hunter, M.G., et al. 2004. 'Endocrine and Paracrine Control of Follicular Development and Ovulation Rate in Farm Species', *Animal Reproductive Science* 82–83: 461–77.

Hussain, A. 1997. *Managing Overweight and Obesity*. Punjagutta: ICFAI University Press.

International Conference on Population and Development (ICPD). 1994. *Programme of Action*. New York: United Nations Population Division, Department for Economic and Social Information and Policy Analysis; available at: http://www.iisd.ca/linkages/Cairo/program/p00000.html.

International Institute for Population Sciences (IIPS) and ORC Macro. 2000. *National Family Health Survey (NFHS-2) 1998–99*: India. Mumbai: IIPS.

Jarow, J.P. 2006. 'Obesity and Erectile Dysfunction', *Journal of Urology* 176(4): 1519–23.

———, et al. 1993. 'Effect of Obesity and Fertility Status on Sex Steroid Levels in Men', *Urology* 42: 171–4.

Jejeebhoy, S. 1998. 'Infertility in India: Levels, Patterns and Consequences Priorities for Social Science Research', *Journal of Family Welfare*, 44: 15–24.

Kadowaki, T., et al. 2003. 'Molecular Mechanism of Insulin Resistance and Obesity', *Experimental Biology and Medicine* 228: 1111–17.

Lake, J.K., C. Power and T.J. Cole. 1997. 'Women's Reproductive Health – the Role of Body Mass Index in Early and Adult Life', *International Journal of Obesity* 21(6): 432–38.

Larsen, U. 2003. 'Infertility in Central Africa', *Tropical Medicine and International Health* 8(4): 354–67.

MacCormack, C.P. 1982. *Ethnography of Fertility and Birth*. London: Academic Press.

Mitchell, M., et al. 2005. 'Adipokines: Implications for Female Fertility and Obesity', *Reproduction* 130: 583–97.

Moran, L.J. and R.J. Norman. 2002. 'The Obese Patient with Infertility: A Practical Approach to Diagnosis and Treatment', *Nutrition in Clinical Care* 5(6): 290–7.

Mulgaonkar, V.B. 2001. *A Research and an Intervention Programme on Women's Reproductive Health in Slums of Mumbai*. Mumbai: Sujeevan Trust.

Norman, R.J, and A.M. Clark. 1998. 'Obesity and Reproductive Disorders: A Review', *Reproduction, Fertility and Development* 10: 55–63.

Popkin, B.M. 1993. 'Nutritional Patterns and Transitions', *Population and Development Review* 19: 138–57.

Riedner, C.E., et al. 2006. 'Central Obesity is an Independent Predictor of Erectile Dysfunction in Older Men', *Journal of Urology* 176: 1519–23.

Schmidt, L. and K. Munster. 1995. 'Infertility, Involuntary Infecundity, and the Seeking of Medical Advice in Industrialized Countries 1970–1992: A Review of Concepts, Measurements and Results', *Human Reproduction* 10: 1407–18.

Scrimshaw, N.S. 1995. 'The New Paradigm of Public Health Nutrition', *American Journal of Public Health* 85: 622–24.

Shetty, P.S. 2002. 'Nutrition Transition in India', *Public Health Nutrition* 5(1A): 175–82.

Thonneau, P. and A. Spira. 1990. 'Prevalence of Infertility: International Data and Problems of Measurement', *European Journal of Obstetrics and Gynaecology Reproductive Biology* 38: 43–52.

Unisa, S. 1999. 'Childlessness in Andhra Pradesh, India: Treatment Seeking and Consequences', *Reproductive Health Matters* 7: 54–64.

Yen, S.S.C, R.B. Jaffe and R.L. Barbieri. 1999. *Reproductive Endocrinology: Physiology, Pathophysiology, and Clinical Management* (fourth edn.). Philadelphia, PA: W.B. Saunders.

——— 1999. 'Chronic Anovulation due to CNS-hypothalamic-pituitary Dysfunction', in S.S.C. Yen, R.B. Jaffe and R.L. Barbieri (eds), *Reproductive Endocrinology: Physiology, Pathophysiology, and Clinical Management* (fourth edn.). Philadelphia, PA: W.B. Saunders, pp. 593–625.

Zurayk, H., et al. 1993. 'Concepts and Measures of Reproductive Morbidity', *Health Transition Review* 3(1): 17–40.

Chapter 11

REPRODUCING INEQUALITIES:

THEORIES AND ETHICS IN DIETETICS

Lucy Aphramor and Jacqui Gingras

Introduction

According to the Director of Policy and Public Affairs of the International Obesity Task Force, 'the suggestion that there is growing 'concern' about the validity of serious health issues associated with obesity is really quite bizarre' (Rigby 2006: 79). That this challenge is purportedly coming from 'academics concerned chiefly with legal, social, political, and educational issues' is also seen as singular. When, as we will demonstrate, 'the decision to feed the world/is the real decision' (Rich 1984: 231), what does this claim for unimpeachable disciplinary authority on fatness signify? As for fatness, food and childbearing, whose business is this?

In this chapter the authors, two feminist dietitian practitioner/ scholars, argue that it is precisely this rich plurality of perspectives that could disrupt the positivistic[1] enterprise that is dietetics by imbuing the field with a feminist science and making visible the more oppressive tendencies that have come to constitute our practice.

Using examples drawn mainly from the United Kingdom and Canada we outline our concern with the way in which, by its heavy reliance on contemporary biomedical discourse to the neglect of engaging with ideas on the social and cultural production of bodies, dietetics' take on obesity unintentionally reproduces health inequities in individuals and communities across the life-course and through generations. As with other authors critical of a lack of analytical rigour in health policy (Saguy and Riley 2005), we use the term anti-obesity to 'signal the moral and political valence' of work premised on the ideological position that fatness should first be diagnosed as the disease obesity and then fought.

In the contemporary context of anti-obesity rhetoric, which is a particularly (and ironically) under-theorised context (Anderson 2005; Aphramor 2005), claims can be made for the constant surveillance and undermining of feeding practices carried out mostly by women in the every day/every night rituals of childbearing and childrearing (Donnelly 2007; Purbrick 2007). Within such a context, focus on 'feeding the family' occludes and, worse, reconstitutes inequities around health. 'Fatness,' or more accurately, 'fear of fatness' narratives that dominate popular culture and the contemporary psychic space of many Westernised nations, eclipse 'real' problems associated with the extraordinary acts of feeding self and Others. This chapter represents an attempt to add theoretical perspective to the anti-fatness morality campaign in the context of childrearing as it plays out within and around women's bodies. The mundane acts of eating in a Western context are localised as a site where fatness is vilified and discord is resoundingly struck among parents who (do or do not) express concern and worry for the future of their child's (and their own) weight and health. In offering a theoretical perspective to this highly particular activity of anti-fat, it is hoped that health professionals (read: dietitians) will entertain a socially integrated feminist dialogue that is justifiably critical towards the moral panic around fatness, women's bodies, and children's eating, a panic in which 'Health is never simply "health" … [but] a means of moralising of normalising and of regulating' (Parr 2002: 373).

Health in every respect: understanding fatness

Contemporary (Westernised) dietetics reflects and amplifies the dominant biomedical preoccupation with reducing/normalizing body size. In teaching, practice and policy, the dietetics field consistently subscribes to the widely held belief in the effectiveness of the concept of 'energy balance' to mould bodies as a means for determining health. Although a range of arguments refuting and contesting the central tenets of the mainstream anti-obesity drive are now increasingly rehearsed (Campos 2004; Gard and Wright 2005; Austin 1999; Blair and Church 2004; Ikeda et al. 1999) the profession has yet to seriously engage with data that problematises the 'obesity problem' (Murray 2005).

Two of these arguments are especially pertinent to this chapter. The first regards the claim that fatness per se is a reliable indicator of poor health status. We are not disputing that fatness can serve as proxy for a range of socio-enviro-economic conditions that predispose people to long term chronic conditions that have become known as 'obesity-related' such as hypertension, diabetes, and coronary heart disease (CHD). These conditions are characterised by increased metabolic risk (IMR) seen in changes in insulin resistance, lipid metabolism, blood pressure and so on. IMR can manifest in people of any weight. That said, the paths to fatness and to IMR may share a common starting point and intersect. For example, people under psychological stress are more susceptible to becoming insulin-resistant, which can contribute to weight gain (Vitaliano et al. 2002; Butler et al. 2002). But it does not necessarily follow that everyone who is fat necessarily has symptoms of IMR.

Given the impact of social position in IMR it is a misguided misnomer to refer to hypertension, diabetes and CHD as arising from 'lifestyle choices' and a cruel irony to suggest: 'If you are overweight, then it's time to think about taking control' (BDA 2006a). For in homogenous Westernised societies these 'diseases of affluence' occur chiefly among the poorer classes (Allender et al. 2006a). And they are not restricted to fat people. Consider for example that 'South Asians living in the U.K. (Indians, Bangladeshis, Pakistanis and Sri Lankans), have a higher premature death rate from CHD than average. The rate is 46 per cent higher for men ... The difference in the death rates between South Asians and the rest of the population is increasing' (BHF 2007: 1, 2) Yet, compared to the general population, rates of fatness are low in South Asians living in the U.K. (Allender et al. 2006b). Clearly, we can see that fatness is not in itself a reliable indicator of health. In fact, health is obviously a more complex process than a weight and/or fatness index can reveal. It is known, for example, that fitness is a significant factor in determining cardio-respiratory health, regardless of body weight (Blair and Church 2004). And while high rates of CHD in South Asians living in the U.K. may be partly explained by levels of physical activity, depression, area of residence, employment and socioeconomic circumstances are also relevant (Allender et al. 2006b).

So, while fatness may indicate or arise from underlying pathology fatness is not necessarily an indicator of poor health. It is biologically possible (although societally contingent) to be fat and healthy. In order to clarify our arguments we will refer to fatness as a descriptor of body weight and to IMR when we are talking about a disease state. Using the term 'obesity' diagnoses (fat) healthy people as unhealthy and thus impedes meaningful discourse.

The second tenet of anti-obesity directives, and one circularly enmeshed in the 'fat is pathological' argument, is the 'energy balance' thesis. In this, thinness is assumed to be desirable for health and readily achievable by cutting current energy (kilocalorie/ kilojoule) intake below energy expenditure. The promise is that 'Cutting calories by 500–600 a day, should lead to a steady weight loss of 0.5–1 kg (1–2lbs) per week … one less 50 calorie plain biscuit could help you lose 5lbs (2.3 kg) in a year – and one extra biscuit means you could gain that in a year!' (BDA 2006b). The empirical evidence undermines this view (Mann et al. 2007; Jain 2005). This means of course, that even if illness, including pregnancy complications, can be linked to fatness, diets are not the solution and indeed have inherent risks.

Nevertheless, the metaphor of energy balance exerts an extraordinary force in determining the trajectory of anti-obesity research and subsequent policy. Efforts to explain secular rates of fatness by energy imbalance, to modify obesogenic environments, to reduce calorie consumption and increase activity levels continue apace despite the observation that data supporting this approach is 'largely circumstantial' (Keith et al. 2006: 2), relying heavily on ecological correlations rather than individual-level epidemiological data or randomised experiments' (Keith et al. 2006: 2). There is a 'less-than-unequivocal evidential basis' (Keith et al. 2006: 2) for such assumptions as the importance of portion size and television viewing in the aetiology of fatness. The energy balance metaphor encapsulates body malleability as infinitely possible within the matrix of a tightly calculated neatness: perhaps it is the apparent facility and reliability of weight loss proffered by this mindset that lends momentum to the medico-aesthetic moralising tone of anti-obesity literature. The justifying framework wherein all individuals are living as equal stakeholders freely engaged in rationally motivated choice – of food, residence, employment etcetera – might be ludicrous but dies hard (subscribing to this 'just world' hypothesis is a recognised feature of people who tend to discriminate against others (Douglas 1995) including on the basis of fatness (Crandall 1994)).

There is also a crucial theoretical/ideological blind spot in the emphasis on weight loss as a clinical goal. In an extraordinary absence of practical reasoning, thinness is persistently conflated with the notion of improved health. Yet fetal and infant development (A2), diet (B1), and physical activity (C2) influence health risk factors much more than the direct and independent effect of fatness (D1) such that:

$$A2 + B1 + C2 >> D1 \text{ (Campos et al. 2005: 81)}$$

Thus, adult health and weight is the cumulative effect of bio-cultural constraints on the (re)production of gendered bodies, constraints that play out on these bodies as situated ideologies around diet, physical activity, body weight and shape. Linear equations bring their own limitations to critical thought in health care but using models for making sense of IMR that integrate rather than ignore relevant evidence seems a prudent step in the right direction. (For a more detailed discussion of the life-course perspective as a framework for the study of nutritional health see Wethington 2005.)

In the grip of Energy Balance, dominant nutritional discourse laments the lack of accurate calorie reporting as a fundamental flaw in (anti-)obesity research (Winkler 2005) while critical social theorists, such as Marsh, simultaneously comment that:

> Unlike the alleged effect of food advertising, the impact of social inequalities on levels of obesity can be measured, and it is very substantial – the largest single factor that has so far been identified. Despite this, it receives scant attention in the media. (Marsh 2007: 8)

Implications for women and children

The argument that social origins have a long-term impact on levels of population fatness (and IMR) builds on critiques (Barker and Osmond 1986; Barker 1998) that highlight the role of maternal and fetal nutrition in influencing long-term health and body weight, including psychological (Bellingham-Young and Adamson-Macedo 2003), ethnocultural, and geographical variables (Lee et al. 2005). Babies born to undernourished mothers are predisposed to IMR and to be fat as adults (Breier et al. 2001). The detrimental effects of poor fetal nourishment can be modified by early infant nutrition. It follows that looking primarily to 'energy balance' to account for, and address, any increased prevalence of IMR will fail: it is not calorie counting but the political transformation necessary to ensure adequate nutrition for all that will make the difference to global patterns in IMR. The American Dietetic Association points out that improved women's education, access to health care and living environment was responsible for 75 per cent of the total reduction in child malnutrition from 1970 to 1995 in sixty-three countries (ADA 2003), a strategy that will concomitantly help prevent the development of IMR when these children mature. And not a slimming plan in sight. Elsewhere, high rates of smoking as a form of weight control among adolescent girls and children (Kendzor et al. 2007; Austin and Gortmaker 2001) has obvious health consequences for themselves and for fetal/adult health of their children.

Complexity theory

So the thinness drive is a liability. The behavioural dysregulation of eating and appetite, generated (in some) by its mores, impacts on early child care. 'Studies of the attachment patterns of parents and their children suggest that early experiences of caregivers may contribute to the intergenerational transmission of physical and psychological vulnerability.' (Brunner 1997: 1472) And how can the Energy Balance mindset help? While the positivistic slant favoured by mainstream dietetics has its place in researching gross nutritional deficiencies, the same reductionist approach is itself grossly deficient when, as is so often the case, it is applied as the dominant valorised approach to research and understanding. Consider, for example, the classic serendipitous observation made during a nutrition trial conducted in two orphanages. Here it was found that the energy content of the children's diet and the type of care they received synergistically influenced growth rate. Children with a caring matron thrived even though receiving a poorer quality diet than children under the authority of a disciplinarian matron (Skuse, Reilly and Wolke 1994). Similarly, zinc supplementation improves growth rate in undernourished children and its effect is enhanced by psychosocial stimulation (Meeks-Gardner 2005). The profile of mainstream anti-obesity narratives provides a metonym for an approach that tends to reify quantitative methodologies and skirt over socially embedded realities: an approach that clearly encumbers any disciplines' contribution to the field of dietetics.

An approach to nutritional health that explicitly attempts to vulgarise and make vulnerable the fat body/person and keep health tightly encapsulated in body weight is a burden to wellbeing. 'Let's focus on why you want to lose weight. It might be to: Look good, Feel better, Live longer' (BDA 2006a: 3) is an example of how 'social structures generate tendencies, sometimes exercised unrealized (as with institutional racism or sizism), and they permit ideological rationalizations that legitimate iniquitous practices that have real effects (Porter 1993)' (in Monaghan 2008). From beyond Energy Balance/mainstream dietetics comes the theory that 'Obesity may simply be a natural consequence of the disease of dieting' (Reto 2003: 148) and that, 'At the very core of the [dieting] dilemma is the fact that we have lost the [intuitive] ability to know how to eat' (Reto 2003: 140). The conscious overriding of hunger contributes to the aetiology of fatness and fuels distress around food (Orbach 2006). This is reason enough to give up on Energy Balance and nurture instead an assenting attentiveness to bodily cues and insist on broader contours for a cartography/choreography of wellness (Satter 2005; Satter 1987; Satter 1983).

Pregnancy and eating distress

Because, as we have seen, poor fetal and infant nutrition predispose to adiposity, women's nutritional status pre- and peri-conception carries significant weight when theorising fatness. When the media, and disappointingly/outrageously some of the medical profession, too, skew accounts of fertility and maternal weight in reinforcing societal bias against fatness and impugning the fitness of fat women as mothers, they assist in the reproduction of stereotype. Murphy and Morgan's response to comments on guidance issued by the British Fertility Society (BFS) illustrated this point:

> The media furore has focussed exclusively on the obesity advice, ignoring the advice about underweight women ... We regret that the selective reporting of the guidelines led to a missed opportunity to raise awareness of the profound effects of eating disorders on reproductive pathology, both on fertility and on the outcome of pregnancy. (Murphy and Morgan 2006: 654)

The BFS report (Kennedy et al. 2006) states that women at both extremes of body weight should be referred for support. Murphy and Morgan (2006) also note that obstetricians frequently fail to ask about women's body weight control practices and have limited knowledge on the impact of eating disorders, a position that could be arrived at from equating health with thinness. Recent research by Aphramor (in press) exploring how fat/size discrimination impacts women seeking medical support reveals the depth and the cost of our reliance on a weight-centred paradigm.

> I would imagine I've got health problems because of what I've done to my body over the years but they wouldn't say that if I was size 10 or 12, they would say I was healthy. The fact that my circulation is shot and I'm always cold and other problems – I do not think they look at that; it's just image: 'well, you look well'. And I have been treated differently by doctors. They do not ask if you are slim – what do you weigh, what is your diet like? It's the first thing they ask if you are overweight. (Aphramor in press)

These discourses clearly establish weight stigmatization, which has disastrous consequences for those engaging with medial personnel who hold strong anti-fat beliefs (Friedman et al. 2005). As articulated by a third participant, a common theme throughout the research was:

> The way in which it is presented is moralistically, it's you're bad being fat, you're bad being overweight. So of course all those messages saying 'all these people who are obese, they need to do something about it' just ends up making you feel crap about yourself because you are obese, because you then just get that public label, about how you are a problem.

Given the fact that we are practising in a milieu that steeps fatness in negative attributes, consistent explicit efforts are needed by health personnel to ensure we counter, rather than unwittingly bolster, this derogation.

The ways in which shifting medico-cultural ideologies around infant feeding practices are taken up and moulded by the media has been shown to have a measurable impact on rates of breastfeeding in the U.S.A. (Foss and Southwell 2006) through a variety of pathways. For instance, it is suggested that 'a mother's prenatal care and the influence of her health professionals, her perception of the father's views on breastfeeding, and her fear of lack of an adequate milk supply for the infant' and that 'the sexualization of the breast in the 1900s, in which mass media outlets played no small role, also may have been a factor in explaining periodic declines in breastfeeding' (Foss and Southwell 2006: 14). A content analysis of infant feeding representations in U.K. media led the authors to conclude that 'Health professionals and policymakers should be aware of patterns in media coverage and the cultural background within which women make decisions about infant feeding' (Henderson et al. 2000: 1196). Breastfeeding was constructed as difficult, embarrassing and class-bound. The picture in Australia was more mixed (Henderson 1999) with breastfeeding contextualised as natural but problematic. And there is a growing literature on the dilemmas of decision making for HIV-positive mothers with limited income and/or access to formula milk and medication.

Moving onwards from fat-blame and energy balance

There are variables that fall outside the energy balance thesis that impact on fatness and maternal reproductive health. Some of these work at the level of dietary composition, which is itself a composite of gendering structural dictates around bodies and childrearing. These dictates have other axes of impact: ethnicity, disability, age, for example. Thus shift work is associated with coronary heart disease and pregnancy complications (Knutsson 2003) and increased rates of late fetal loss (Zhu et al. 2004). And our place in the social hierarchy makes an impress on body weight regardless of our 'chosen' calorie intake/expenditure via metabolic pathways in which sleep debt and pharmaceutical iatrogenesis (Keith et al. 2006), oppression (Butler et al. 2002) and psychological distress (Vitaliano et al. 2003) contribute to fatness independent of any effects on energy intake or expenditure.

Dramaturgical stress, which arises from being continuously required to repress anger in the face of the arbitrary exercise of power, or needing to maintain face inconsistent with your sense of self – the self-parodying fat person, the queer woman who fears hate crime, the black person whose anger is invalidated – is associated with hypertension and IMR (Freund and McGuire 1995). Not surprisingly, IMR, also known as oppression syndrome, has been described as a stress-related disorder (Tappy 2004). Explaining any weight-linked differences in health status simply by fatness, and then by a fat person's failure to lose weight, would be like suggesting that the failure of poor people to join gyms explains social-class differences in morbidity and mortality.

Sociology and geography bring much needed critiques to the popular medically derived belief that fatness is abhorrent, self-determined, and socially costly. The medicalisation of bodies is enabled through Cartesian dichotomizing such that 'bodies are nothing more than automatons, machines acting as containers for the non-spatial mind' (Patterson and Elliott 2002: 231). The objective, empiricist view has had particularly dehumanizing impacts on human dimensions of dietetic practice (Aphramor 2005; Aphramor and Gingras 2008; Gingras 2005). Given the role of dietitians in promoting healthy eating among mothers and children, we concede that our medicalising efforts are complicit in reproducing inequities and occluding the social/relational influences on food, feeding, and eating. Like Evans, we believe that 'in addition to questioning the certainty of "scientific" knowledge on obesity, understanding the context for the production, reproduction and circulation of knowledge about obesity (and fatness in general) is central to the development of such critical approaches' (2006: 260). If not for these critical transdisciplinary conversations in which we now engage, unethical, invisible encounters in the name of nutritional health would go unnoticed, as do many of our professional contributions to the field of public health as perceived by colleagues in allied health professions (Devine, Jastran and Bisogni 2004). Reconciling such significant discrepancies holds promise for the sustainable, collective, and respectful effort of the dietetic profession. But as Jardine notes, 'There is both sadness and adventure ahead, and there is pain to pay for the somnambulant beliefs in our own dominion' (1998: 135), beliefs whose authority has led to the eclipsing of fat narratives and thus, creation of nutrition inequities in dietetic practice designed to regulate the childbearing, childrearing female body.

We write this as women, partners, mothers, sisters, aunts, and daughters. We live the contradictions of theory and practice in the messy, always emotional praxis of feeding self and child(ren). We have experienced the surveillance of the medical gaze on our bodies, pregnant or otherwise, and were left feeling vulnerable and shamed. The [pregnant] female body, rife with and representative of uncertainty, ambiguity, and conflicts of knowledge, resists biomedical science, but in most cases becomes 'obfuscated and very difficult to see … once it enters the public domain' (Evans 2003: 88). The [pregnant] body is positioned as an oversimplified entity in the nutrition discourse domain, which tends to conceive of all bodies 'as corrupt and flawed, requiring the liberatory intervention of rationality acting through science and technology' (Patterson and Elliott 2002: 231). As Bryn Austin notes, 'nutritional public health has demonstrated little understanding of anything outside of traditional materialist ideology', resulting in a systematically over-defined bedrock for the profession, one that crucially maintains 'nutritional science's incognizance of its relationship to experiences of eating, dieting, and body image' (1999: 246) and, we would add, fat prejudice. In dietetics, we are more than happy to comply by offering such an intervention and marking 'non-compliant' those bodies that continue to resist our efforts. As the discontent grows within and beyond our profession (Devine, Jastran and Bisogni 2004), there is a call for a renewed ethic of food and nutrition practice. One of the first sites for subversion is the classroom in which dietetic students are educated.

Arriving late: lags in dietetic education

While there is no explicit curricular theory framing dietetic education, critical discourse analysis reveals that within dietetic education, learning processes are sequential and apolitical and dietetic knowledge is decontextualised from the social world (Gingras, under review). This professionalization process results in a dietetic knowing that is partial, arrested, and fragmentary. By inference, dietetic education is constituted by structuralist theories of behavioural objectivism. This theoretical framework assumes curriculum-as-plan, where competence is associated with efficiency-based, instrumental action. Such a view neglects 'what matters deeply in the situated world of the classroom – how the teachers' 'doings' flow from who they are ... Teaching is fundamentally a mode of being' (Aoki 1986: 8). The curriculum is dehumanised, which permits omission of the social aspects of dietetic practice from the curriculum. Gould suggests, 'Science, since people must do it, is a socially embedded activity' (1981: 22).

In the context of learning about the dietetic professional's 'role', dietetic students are (mostly) not encouraged to consider anything but the 'fat as pathological' biomedical, anti-social-science discourse. There is a clear message to students that the professional is the authority over the body. Through behavioural objectivism and overly wrought and under-theorised learning competencies on which dietetic education is traditionally informed, a rampant decentring occurs; the student is decentred from her own body as she learns to decentre the Other since the 'educational value is authoritatively located in structures external to individuals, and it is ideologically neutral' (Cherryholmes 1988: 30). Berry considers the effects of professional competencies/standards on professional education as that which:

> explains the decline in education from ideals of service and citizenship to mere 'job training' or 'career preparation.' The religion of professionalism is progress, and ... in spite of its vocal bias in favour of practicality and realism, professionalism forsakes both past and present in favour of the future, which is never present or practical or real. Professionalism is always offering up the past and the present as sacrifices to the future, in which all our problems will be solved and our tears wiped away – and which, being the future, never arrives. (2000: 131)

These issues should be of great concern to a profession such as dietetics, which can have a profound collaborative influence on reducing nutrition inequities and making visible the female childbearing body. Ellsworth claims that education is where the social construction of knowledge and learning gets deeply personal. She explains that the pedagogical 'mode of address is one of the those intimate relations of social and cultural power ... whose subtleties can shape and misshape lives, passions for learning, and broader social dynamics' (1997: 6). The workings of power influence how dieticians are educated because the 'mode of address' is aimed at:

> shaping, anticipating, meeting, or changing who a [dietetic] student thinks [s/he] is ... in relation to gender, race, sexuality, social status, ability, religion, ethnicity, and all those other differences that, at this historical moment, are used to make a difference in opportunity, health care, safety, sense of self, employment, quality of life. (Ellsworth 1997: 7)

This process has gone virtually unexplored in dietetics education. The 'mode' in which dietetic students are 'addressed' by curriculum, text, instructors, professional associations, and regulatory bodies has immediate implications for their personal and professional lives. Good (1994) in speaking of medical education asks 'Can we seriously contemplate an epistemological – and ethical – stance that does not privilege the knowledge claims of biomedicine and the biomedical sciences?' We would contend that we can contemplate such a stance and that in the midst of a crisis of professionalism,[2] it is crucial to reflect on all possibilities for dietetic education, especially that which is made possible through an integrated feminist science. Campbell in theorising feminist science, suggests that integrating 'Haraway's work represents a rich and provocative possibility for conceiving new strategies for producing reflexive feminist science studies' (2004: 163). In this way, feminist science studies could openly acknowledge dietetics as socially constituted:

> [A] notion of strong reflexivity would require that the objects of inquiry be conceptualised as gazing back in all their cultural particularity and that the researcher, through theory and methods, stand behind them, gazing back at [her] own socially situated research project in all its cultural particularity and its relationships to other projects of [her] culture. (Harding 1991: 163)

We ask this of dietetic knowledge production, especially given that structuralist discourses of dietetic curriculum have the potential to institutionalise discriminatory, oppressive practices. Unexamined educational processes function as a technology of regulation and surveillance (Lather 1991). A reflexive, critical appraisal and renegotiation of what counts as dietetic education and epistemology is urgently needed.

Given the relatively static, one-hundred-year history of dietetic curriculum as that which is founded on principles of objectivity and rationality, the implications for how knowledge is translated into dietetic practice may be equally troubling. We understand, but do not condone, addressing mothers and/or children as essentialised entities into which dietitians pour nutrition rules, formulas, and behaviour change theories, then respond with mother-blame when these bodies go seemingly awry or more offensively, do not comply (Lepage, Moisan and Gaudet 2004). If our current dietetic curriculum is based on behavioural objectivism, mastering learning, competencies, and knowledge statements and those approaches are considered unethical, then those who are charged with educating dietitians may be moved to review accreditation processes and encourage educators and professional associations to institute alternative educational experiences.

Even though dietetic curricular theory has persisted in the tradition of positivism for over a century, 'it is possible to teach [dietetic] students [in a way] that doesn't require them to assume a fixed, singular, unified position within power and social relations' (Ellsworth 1997: 9). With careful attention to mode of address, imagination, and a socially integrative feminist science (Haraway 1991) reflected by significant (re)imagination and (re)vision to the dietetic curriculum, we might observe a most remarkable shift in dietetic education, which has implications for student engagement, dietetic epistemology, and practice. A comparable call to action was made over a decade ago, yet little movement has been observed (McDonald et al. 1993). As health profession educators, this educational possibility is within our reach as we consider more profoundly this moment within a collective historical biography.

Conclusion

In this chapter we have explored how new perspectives could disrupt dietetics as a positivistic enterprise, bringing more of our lives into view. Becoming aware of our particular social positions in a dominant educational and practice culture carries with it a sense of belatedness; we become conscious of the weight of our history, consciousness bearing a reminder that 'we are particular kinds of people, rather than unique and unprecedented individuals' (Levinson 2001: 15). This awareness is crucial for us to begin to articulate the educational possibility of a poststructural, socially just, feminist dietetic education and practice. However, if our belatedness paralyses us, if we are unable to process the shock of the old, we cannot act and incorporate the new: the status quo remains intact, always lagging behind. We are left with the question: how might we encourage and preserve newness in ways that foster our understanding of food and nutrition that does not reproduce inequities? As Arendt (1968) argues, it is only in assuming an active relation to our world, one that embraces what has been occluded, that we can bring to it the fresh perspectives needed to reform our educational and practice encounters.

Notes

1. Positivism is a philosophical movement that holds that all meaningful statements are conclusively verifiable through observation and experiment and that metaphysical theories are therefore strictly meaningless. Positivism is a philosophical orientation that considers theoretical knowledge as value-free, objective, and unrelated to practice. The positivistic tradition is historically associated with scientific processes and technical rationality (Vaines 1997).
2. Hafferty (2000) concludes it is the rise of the technician that is under attack (in crisis) since the professional-as-technician is unable to address larger social transformation by stifling the moral imagination. Richardson

(1997) describes the crises as uncertainty about what constitutes adequate depiction of social reality.

References

Allender, S., et al. 2006a. 'Prevalence of Obesity by Sex and Ethnic Group, 2004, England', in *Diet, Physical Activity and Obesity Statistics*. London: British Heart Foundation, figure 3.8, p. 81; available at: http://www. heartstats.org.

————, et al. 2006b. 'Prevalence of Overweight and Obesity by Sex and Socio-economic Classification, 2003, England and Scotland, 2003/04, Wales', in *Diet, Physical Activity and Obesity Statistics*. London: British Heart Foundation, figure 3.7, p. 80; available at: http://www.heartstats.org.

Anderson, A. 2005. 'Obesity Prevention and Management – Evidence and Policy', (editorial) *Journal of Human Nutrition and Dietetics* 18: 1–2.

Aoki, T. 1986. 'Teaching as In-dwelling between Two Curriculum Worlds', *The BC Teacher* April/May: 8–10.

Aphramor, L. 2005. 'Is a Weight-centred Health Framework Salutogenic? Some Thoughts on Unhinging Certain Dietary Ideologies', *Social Theory and Health* 3: 315–40.

———— and J.R. Gingras. 2008. 'Sustaining Imbalance: Neglecting the Evidence in Pursuit of Nutritional Health', in S. Riley, et al. (eds), *Critical Bodies: Representations, Practices and Identities of Weight and Body Management*. London: Palgrave/Macmillan, pp. 155–74.

———— and J.R. Gingras. 2009. 'That Remains to be Seen: Disappeared Feminist Discourses on Fat in Dietetic Theory and Practice', in E.D. Rothblum and S. Solovay (eds), *The Fat Studies Reader*. New York: New York University Press, pp. 97–105.

———— in press. 'The Impact of a Weight-centred Treatment Approach on Women's Health and Health-seeking Behaviours', *Journal of Critical Dietetics*.

Arendt, H. 1968. 'The Crises in Education', in *Between Past and Future*. New York: Viking Press, pp. 173–96.

Austin, S.B. 1999. 'Fat, Loathing and Public Health: The Complicity of Science in a Culture of Disordered Eating', *Culture, Medicine and Psychiatry* 23: 245–68.

Austin S.B. and S.L. Gortmaker. 2001. 'Dieting and Smoking Initiation in Early Adolescent Girls and Boys: A Prospective Study', *American Journal of Public Health* 91(3): 446–50.

Barker, D.J. 1998. *Mothers, Babies and Health in Later Life*. Edinburgh: Churchill Livingstone.

———— and C. Osmond. 1986. 'Infant Mortality, Childhood Nutrition and Ischaemic Heart Disease in England and Wales', *The Lancet*, 1: 1077–81.

Bellingham-Young, A. and E.N. Adamson-Macedo. 2003. 'Foetal Origins Theory: Links with Adult Depression and General Self-efficacy', *Neuroendocrinology Letters* 6(24): 412–16.

Berry, W. 2000. *Life Is a Miracle: An Essay against Modern Superstition*. Washington, DC: Counterpoint.

Blair, S.N. and T.S. Church. 2004. 'The Fitness, Obesity and Health Equation: Is Physical Activity the Common Denominator?', *Journal of the American Medical Association* 292: 1232–34.

Breier, B.H., et al. 2001. 'Fetal Programming of Appetite and Obesity', *Molecular and Cellular Endocrinology* 185: 73–79.

British Dietetic Association (BDA). 2002. 'Be Weight Wise: Why Being a Healthy Weight Is So Important', *Fact Sheet* 1.

———— 2004. *Size Matters Campaign Literature*. London: BDA.

———— 2006a. 'R U Ready? Weight Wise', accessed 17 April 2007 at: http://www.bdaweightwise.com/ready.aspx.

———— 2006b. 'FAQ. Weight Wise', accessed 17 April 2007 at: http://www.bdaweightwise.com/expert/expert_faq.aspx#top.

British Heart Foundation (BHF). 2007. 'Ethnic Differences in Mortality', accessed 4 April 2007 at: http://www.heartstats.org/datapage.asp?id=737.

Butler, C., et al. 2002. 'Internalized Racism, Body Fat Distribution, and Abnormal Fasting Glucose among African-Caribbean Women in Dominica, West Indies', *Journal of the National Medical Association* 94(3): 143–48.

Calling, S. 2006. 'Obesity and Cardiovascular Disease: Aspects of Methods and Susceptibility'. Ph.D. disseration. Malmo: Lund University; published on the Internet; accessed 20 November 2006 at: http://theses.lub.lu.se/postgrad//search.tkl?field_query.

Campos, P. 2004. *The Obesity Myth: Why America's Obsession with Weight Is Hazardous to Your Health*. New York: Gotham Books.

————, et al. 2005. 'Response: Lifestyle Not Weight Should Be the Primary Target', *International Journal of Epidemiology* 35: 81–82.

Chandola, T., E. Brunner and M. Marmot. 2006. 'Chronic Stress at Work and the Metabolic Syndrome: Prospective Study', *British Medical Journal* 332: 521–5.

————, et al. 2004. 'The Effect of Control at Home on CHD Events in the Whitehall II Study: Gender Differences in Psychosocial Domestic Pathways to Social Inequalities in CHD', *Social Science and Medicine* 58: 1501–09.

Cherryholmes, C. 1988. *Power and Criticism: Poststructural Investigations in Education*. New York: Teachers College Press.

Crandall, C.S. 1994. 'Prejudice against Fat People: Ideology and Self-interest', *Journal of Personal and Social Psychology* 66: 882–94.

———— and R. Martinez. 1996. 'Culture, Ideology and Anti-fat Attitudes', *Personality and Social Psychology Bulletin* 22(11): 1165–68.

Devine, C. M., M. Jastran and C.A. Bisogni. 2004. 'On the Front Line: Practice Satisfactions and Challenges Experienced by Dietetics and Nutrition Professionals Working in Community Settings in New York State', *Journal of the American Dietetic Association* 104(5): 787–92.

Dietz, W.H. 1998. 'Health Consequences of Obesity in Youth: Childhood Predictors of Adult Disease', *Pediatrics* 101: 518–25.

Donnelly, N. 2007. 'Editorial', *Network Health DietitiansDieticians* 22: 3.

Douglas, T. 1995. *Scapegoats: Transferring Blame.* London: Routledge.

Ellsworth, E. 1997. *Teaching Positions: Difference, Pedagogy, and the Power of Address.* New York: Teachers College Press.

Evans, B. 2006. ' "Gluttony or Sloth": Critical Geographies of Bodies and Morality in (Anti)obesity Policy', *Area* 38(3): 259–67.

Foss, K.A. and B.G. Southwell. 2006. 'Infant Feeding and the Media: The Relationship between *Parents' Magazine* Content and Breastfeeding, 1972–2000', *International Breastfeeding Journal* 1: 10.

Friedman, K.E, et al. 2005. 'Weight Stigmatization and Ideological Beliefs: Relation to Psychological Functioning in Obese Adults', *Obesity Research* 13(5): 907–16.

Gard, M. and J. Wright. 2005. *The Obesity Epidemic: Science, Morality and Ideology.* London: Routledge.

Gingras, J.R. 2005. 'Evoking Trust in the Nutrition Counsellor: Why Should We Be Trusted?', *Journal of Agricultural and Environmental Ethics* 18: 57–74.

———— 2009 'The Educational (Im)possibility for Dietetics: A Poststructural Content Analysis of the *Accreditation Manual*', *Learning Inquiry* 3(3):177-191.

Good, B.J. 1994. *Medicine, Rationality, and Experience.* Cambridge: Cambridge University Press.

Gould, S.J. 1981. *The Mismeasure of Man.* New York: Norton.

Hafferty, F.W. 2000. 'In Search of a Lost Chord: Professionalism and Medical Education's Hidden Curriculum', in D. Wear and J. Bickel (eds), *Educating for Professionalism: Creating a Culture of Humanism in Medical Education.* Iowa City: University of Iowa Press, IAss, pp. 11–34.

Haraway, D. 1991. *Simians, Cyborgs, and Women: The Reinvention of Nature.* London: Free Association.

Harding, S. 1991. *Whose Science? Whose Knowledge? Thinking from Women's Lives.* Milton Keynes: Open University Press.

Henderson, A. 1999. 'Mixed Messages about the Meaning of Breast-feeding Representations in the Australian Press and Popular Magazines', *Midwifery* 15(1): 24–31.

Henderson, L., J. Kitzinger and J. Green. 2000. 'Representing Infant Feeding: Content Analysis of British Media Portrayals of Bottle Feeding and Breast Feeding', *British Medical Journal* 321: 1196–98.

Ikeda, J., et al. 1999. 'A Commentary on the New Obesity Guidelines from NIH', *Journal of the American Dietetic Association* 99: 918–20.

Jain, A. 2005. 'Treating Obesity in Individuals and Populations', *British Medical Journal* 331: 1387–90.

James, W.P., et al. 1997. 'Socioeconomic Determinants of Health: The Contribution of Nutrition to Inequalities in Health', *British Medical Journal* 314: 1545–49.

Jardine, D. 1998. *To Dwell with a Boundless Heart: Essays in Curriculum Theory, Hermeneutics, and the Ecological Imagination*. New York: Peter Lang.

Keith, S.W., et al. 2006. 'Putative Contributors to the Secular Increase in Obesity: Exploring the Roads less Travelled', *International Journal of Obesity* 30: 1585–94.

Kendzor, D.E., et al. 2007. 'Weight-related Concerns Associated with Smoking in Young Children', *Addictive Behaviors* 32: 598–607.

Kennedy, R., et al. 2006. 'Implementation of the NICE Guideline – Recommendations from the British Fertility Society for National Criteria for NHS Funding of Assisted Conception', *Human Fertility* 9(3): 181–89.

Knutsson, A. 2003. 'Health Disorders of Shift Workers', *Occupational Medicine* 53: 103–108

Lather, P. 1991. *Getting Smart: Feminist Research and Pedagogy with/in the Postmodern*. New York: Routledge.

Lepage, M-C., J. Moisan and M. Gaudet. 2004. 'What do Québec Children Eat during their First Six Months?', *Canadian Journal of Dietetic Practice and Research* 65(3): 106–13.

Levine, M. and L. Smolak. 1998. 'The Mass Media and Disordered Eating', in W. Vandereycken and G. Noordenbos (eds), *Prevention of Eating Disorders*. London: Athlone Press.

Levinson, N. 2001. 'The Paradox of Natality: Teaching in the Midst of Belatedness', in M. Gordon (ed.), *Hannah Arendt and Education: Renewing our Common World*. Boulder, CO: Westview Press, pp. 11–36.

McDonald, B.E., et al. 1993. 'Concept of Dietetic Practice and Framework for Undergraduate Education for the 21st Century', *Journal of the Canadian Dietetic Association* 54(2): 75–80.

Monaghan, L. 2008. *Men and the War on Obesity: A Sociological Study*. London: Routledge.

Murphy, H.C. and J.F. Morgan. 2006. 'Society's Advice on Low Weight and IVF Was Ignored by the Media', (Letter) *British Medical Journal* 333: 654.

Murray, S. 2005. 'Introduction to "Thinking Fat" ', *Special Issue, Social Semiotics* 15(2): 111–12.

National Statistics. 2007. 'Spreadsheet ST340712 Obesity among Adults: By Sex and NS-SeC, 2001: Social Trends 34 Updated 2004', in P. Marsh *Poverty and Obesity*; accessed 4 April 2007 at: http://www.sirc.org/articles/poverty_and_obesity.shtml.

Olson, C.M. and W.S. Wolfe. 2005. 'Parity and Body Weight in the United States: Differences by Race and Size of Place of Residence', *Obesity Research* 13: 1263–69.

Parr, H. 2002. 'Medical Geography: Diagnosing the Body in Medical and Health Geography, 1999–2000', *Progress in Human Geography*, 26: 240–51.

Patterson, M. and R. Elliott. 2002. 'Negotiating Masculinities: Advertising and the Negotiation of the Male Gaze', *Consumption, Markets and Culture* 5: 231–46.

Purbrick, T. 2007. 'Parade of the Podge: Kids Need to be Brought Up, Not Buttered Up', *Network Health Dietitians* 22: 14.

Rich, A. 1984. 'Hunger' *in The Fact of A Doorframe*. New York: W.W. Norton, p.229.

Richardson, L. 1997. *Fields of Play: Constructing an Academic Life*. New Brunswick, NJ: Rutgers University Press.

Saguy, A.C. and K.W. Riley. 2005. Weighing Both Sides: Morality, Mortality, and Framing Contests over Obesity', *Journal of Health Politics, Policy and Law* 30(5): 875.

Satter, E. 1983. *Child of Mine: Feeding with Love and Good Sense*. Palo Alto, CA: Bull Publishing Company.

———— 1987. *How to Get Your Kid to Eat, but Not Too Much*. Palo Alto, CA: Bull Publishing.

———— 2005. *Your Child's Weight: Helping without Harming*. Madison, WI: Kelcy Press.

Skuse, D., S. Reilly and W. Wolke. 1994. 'Psychosocial Adversity and Growth during Infancy', *European Journal of Clinical Nutrition* 48(Suppl 1): S113–30.

Vaines, E. 1997. 'Professional Practice and Families: Searching for Maps to Guide Ethical Action', in J. F. Laster and R. G. Thomas (eds), *Thinking for Ethical Action in Families and Communities, Seventeenth Yearbook of Education and Technology Division, American Association of Family and Consumer Sciences*. Peoria, IL: Glencoe/McGraw-Hill, pp. 203–16.

Vitaliano, P.P., et al. 2002. 'A Path Model of Chronic Stress, the Metabolic Syndrome, and Coronary Heart Disease', *Psychosomatic Medicine* 64: 418–35.

Wethington, E. 2005. 'An Overview of the Life Course Perspective: Implications for Health and Nutrition', *Journal of Nutrition Education and Behavior* 37: 115–20.

Winkle, J.T. 2005. 'The Fundamental Flaw in Obesity Research', *Obesity Reviews* 6(3): 199–202.

Zhu, J.L., et al. 2004. 'Shift Work, Job Stress, and Late Fetal Loss: The National Birth Cohort in Denmark', *Journal of Occupational and Environmental Medicine* 46(11): 1144–49.

NOTES ON CONTRIBUTORS

Lucy Aphramor, R.D., is a dietitian with the social enterprise Atrium Health Ltd. and in the National Health Service, and is an honorary research fellow at Coventry University. Lucy's work seeks to advance a socially integrated approach to improving wellbeing and reducing inequalities. She has published work on critical dietetics, frequently in collaboration with Jacqui Gingras.

Shauna Clarke completed a B.A. and M.A. in Anthropology from the National University of Ireland, Maynooth, where her research focused on nutritional anthropology. She is presently completing a B.Sc. in Nutrition and Dietetics in the University of Chester while continuing with her research interests in the social elements of nutrition and the education of dieticians.

Michael Davies is an Associate Professor and Co-Director of the Research Centre of Early Origins of Health and Disease at the University of Adelaide. His research in reproductive epidemiology investigates early life factors that impact on the health of Australians within and across generations. He is also researching decision-making, safety, and effectiveness of assisted reproductive technologies.

Ama de-Graft Aikins is a Research Fellow in Social Psychology at the University of Cambridge. Her research focuses on experiences of chronic physical and mental illnesses among African communities on the continent and in the U.K.

Jacqui Gingras, Ph.D., R.D., is an Assistant Professor at Ryerson University's School of Nutrition. Her research involves theoretical and experiential explorations of critical health and nutrition epistemology. She is the co-author of several chapters with Lucy

Aphramor that delineate a critical socio-political perspective of dietetic education and practice.

Aravinda Meera Guntupalli is a Lecturer at the University of Southampton. She is a trained anthropologist, demographer and economist and applies her interdisciplinary skills to research on health, gender and inequality in India and the U.K. She has published several articles on anthropometry, gender inequality and reproductive health using qualitative and quantitative techniques.

Nicola Heslehurst BSc, MSc, PhD, is a Senior Lecturer in Research at Teesside University at the Health and Social Care Research Institute, and a member of the North East Maternal Obesity Research Group. Her specialist research area is maternal obesity, specifically relating to the development and improvement of NHS maternity services to support obese women throughout their pregnancy and beyond. Nicola has been involved with the Centre for Maternal and Child Enquiries (CMACE) in their External Advisory Group and Consensus of Standards group for the Obesity in Pregnancy project, which has developed national standards of care for obese women in pregnancy and postnatally.

Mara Mabilia has a Ph.D. in Social Anthropology and is Professor of Cultural Anthropology at the University of Padua. She also lectures on other postgraduate programmes at Universities of Padua and Brescia. She has previously carried out research in Kenya, Tanzania and Italy, and is currently doing field research on female genital mutilation among groups of immigrant women in some Italian regions. She is an Associate Member of the Fertility and Reproduction Studies Group, Institute of Social and Cultural Anthropology, University of Oxford. She has authored numerous scientific publications.

Vivienne Moore is an Associate Professor and (with Michael Davies) is Co-Director of the Life Course and Intergenerational Health Research Group at the University of Adelaide, South Australia. Her research investigates the social and behavioural influences on women's and children's health within a life-course perspective, and gender inequalities in health.

Sara Randall is a Professor at University College London Department of Anthropology. As an anthropological demographer, she researches issues around reproductive decision-making and the demographic consequences of mobility, with a particular focus on francophone West Africa.

Devi Sridhar is a postdoctoral fellow in politics at All Souls College, Oxford and also directs the global health governance project at the Global Economic Governance Programme. She is visiting faculty at the Public Health Foundation of India/Indian Institute of Public Health.

Soraya Tremayne is a social anthropologist and the Founding Director of the Fertility and Reproduction Studies Group and a Research Associate at the Institute of Social and Cultural Anthropology, University of Oxford. Previous to this, she was the Director of the Centre for Cross-Cultural Research, Queen Elizabeth House, University of Oxford. For the past twelve years, she has carried out research on reproduction and sexuality in Iran. Her current research focuses on assisted reproductive technologies in Iran.

Maya Unnithan-Kumar is Reader in Social Anthropology at the University of Sussex. Maya's current research specialism is in medical anthropology, especially in the areas of reproduction, health, gender and the body. Since 1998, she has conducted fieldwork in the district and city of Jaipur on poor women's access to reproductive healthcare. Prior to this, from 1986, she worked on kinship, gender and development issues in southern Rajasthan. Maya is currently leading an Economic and Social Science Research Council (U.K.)-funded research project on NGO engagement with human rights discourse in the fields of sexual, maternal and reproductive health in India.

Saskia Walentowitz is a research fellow at the Institute of Social Anthropology at Bern University. Her main fields of interest are gender, kinship and reproduction in Muslim contexts, as well as infant feeding, and the anthropology of science. She recently co-edited the Social Science & Medicine special issue on 'Women, Mothers and AIDS Care in Resource-Poor Settings'.

Megan Warin is a social anthropologist and senior lecturer in the Discipline of Gender, Work and Social Inquiry at the University of Adelaide. Her current research interests span theories of embodiment and phenomenology amongst Persian women in Australia, and intersections of class and gender in experiences of obesity.

INDEX